Life Sciences Research to Product Development

The present volume, *Life Sciences Research to Product Development: Regulatory Requirement Transforming, Volume 1*, discusses the procedures of drug approval and regulatory requirements that must be met according to the United States Food and Drug Administration (US FDA), the European Medical Agency (EMA), and the Central Drug Standard Control Organization (CDSCO). Many researchers either abandon their work in the middle of the process or find it difficult to follow the rules. Therefore, it is not surprising that any biological researcher associated with drug development should have a thorough understanding of regulatory requirements. This volume incorporates all the requisite regulatory norms and provides the latest information on the mandated regulation of herbal medicines.

The book covers other obligatory regulatory requirements such as:

- The legal method and practice of herbal drug products, the roles of Ayurvedic medicines, and the process to obtain regulatory approval.
- Drug molecules not included in Department of Ayurveda, Yoga, Naturopathy, Unani, Siddha, and homeopathy (AYUSH) but referred to as phytopharmaceuticals are also considered new drugs.
- The boundary line between food and herbal pharmaceuticals is discussed, as well as pre-clinical toxicity testing, clinical trials, and stability studies in accordance with the rules.

The chapter on regulatory implications for the approval process in this book will be the most useful resource for researchers and students, particularly those with backgrounds in pharma, forensic medicine, or regulatory affairs, or those who aspire to succeed in drug research. Additionally, the information contained in this volume of the book could be of great interest to researchers working in the herbal drug industry.

Life Sciences Research to Product Development

Regulatory Requirement Transforming, Volume 1

Pronobesh Chattopadhyay and
Danswrang Goyary

CRC Press is an imprint of the
Taylor & Francis Group, an **informa** business

First edition published 2024.
by CRC Press
2385 NW Executive Center Drive, Suite 320, Boca Raton FL 33431

and by CRC Press
4 Park Square, Milton Park, Abingdon, Oxon, OX14 4RN

CRC Press is an imprint of Taylor & Francis Group, LLC

ISBN: 978-1-032-50426-1 (hbk)
ISBN: 978-1-032-50430-8 (pbk)
ISBN: 978-1-003-39843-1 (ebk)

DOI: 10.1201/9781003398431

Typeset in Times LT Std
by Apex CoVantage, LLC

Contents

Author Biographies

Dr. Pronobesh Chattopadhyay, M.E (Biomedical Engg.), M.Pharm., Ph.D. is a pharmaceutical scientist in DRDO. His working areas are chemical pharmacology, toxicological interaction on expression, chemical target to TRPV, AdipoR1, AdipoR2, c-Jun kinase pathway in lungs and eyes, and these resulted in exploration to development of many defense products. He was instrumental in developing many defense products which are successfully used by India's armed forces. A notable technology like the "chili grenade" developed from *Bhot Jolakia* (one of the hottest chilies in the world) has been transferred (transfer of technology) to three industries and are presently used successfully by many law enforcing agencies in India. He also developed protective gear for ultra-violet protection which are stable at –20°C. He is a Fellow of the National Academy of Sciences India (NASI) and Fellow of Indian Chemical Society (FICS). He has published more than 170 research papers and five book chapters and filed 15 patents while three patents are granted. He has received many awards, including the Gandhi Young Technology Innovation Awards under Biotechnology Industry Research Assistance Council (BIRAC), Laboratory Scientist Awards, Technology Awards, etc. He is a regular reviewer of ethnopharmacology, toxicology, and biochemical pharmacology, and Elsevier publishing house recognized him as a potential reviewer. Twelve Ph.D. degrees and three post doctorate degrees have been awarded under his supervision, and 25 postgraduate students completed their dissertations.

Dr. Danswrang Goyary graduated in Biotechnology and did his M.Sc. and Ph.D. degrees in Biochemistry from Northeastern Hill University (NEHU), Shillong, India. He qualified NET-JRF (UG-CSIR) in 2002 and received a fellowship for his doctoral degree. He started his professional career at the Defence Institute of Bio-Energy Research, Haldwani, Uttarakhand as a scientist in the year 2005 and extensively contributed to the fields of life sciences and biopharmaceuticals of defense importance as a biochemist. He is presently working in Defence Research Laboratory, DRDO, Tezpur (Assam). He is actively involved in studying the mechanism of different kind of wound healing, hemostasis and associated with the development of defense products viz. chili grenades, snake repellent, high-SPF sunscreen, impregnated fabrics for disease vector and insect control, etc. He has contributed more than 50 publications in peer reviewed journals, secured two patents, supervised three Ph.D. students, and guided more than 15 post-graduate students. He has received laboratory awards such as Technology Group awards, Science Day oration award, Technology Day oration award, etc. He is a member of many committees such as institutional animal ethical committees (IAEC) and institutional biosafety committees (IBSC), etc.

Acknowledgments

At the very outset, the authors prayed to almighty God for enlightenment, strength, and knowledge for writing this book and express their heartfelt gratitude to their loved ones, parents, spouses, and children, who provided unwavering encouragement and support throughout the writing of this book. The authors would like to acknowledge to all senior officials of DRDO for their relentless support to write this book on life sciences product development and managing its regulatory processes.

The authors are also thankful to the Director, DRL, Tezpur, Assam for providing space and opportunity to write this book and the colleagues for their ever-ready help and support in acquisition of practical and technical know-how of regulatory processes which contributed to the successful completion of writing this book.

The authors would also like to express their deep gratitude to all students for their help, especially Mr. Mohit Kumar, JRF for his consistent assistance in manuscript formatting.

The authors are indebted to their research collaborators for their unflinching trust, infinite affection, and uncompromising honesty, which were major driving forces behind the writing of these volumes. They are deeply grateful for the constructive suggestions received during various project meetings and reviews which inspired them to create these volumes with sense of purpose, duty towards society at large, and hope for the future.

The authors gratefully acknowledge to Dr. Renu Upadhyay, Ms. Jyotsna Jangra and CRC Press for their patience, guidance, and ever-ready help in making the successful preparation of this book possible. The author also acknowledge the help of Mr. Jayachandran Rajendiran, of Apex Co-Vantage to comment this book.

In the end, the authors express their heartfelt gratitude to their most treasured readers.

Pronobesh Chattopadhyay, Ph.D., FNASc., FICS
Danswrang Goyary, Ph.D.

Preface

The development of innovative drugs adheres to a methodical and rigorous scientific technique. To get insights into the effectiveness and safety of the compounds, they were initially created using computer-based virtual modeling and subsequently synthesized in laboratory test tubes. Following the discovery of a promising therapeutic compound, it is imperative to conduct pre-clinical investigations to assess its efficacy and safety as a potential treatment. Amidst the COVID-19 pandemic, we have observed firsthand the expeditious progress in vaccine research, showcasing the nation's prowess as a prominent global player in the field of medicines. The arduous progression commences with the exploration of pharmaceutical and innovative therapeutic options, followed by clinical investigation, registration, and ultimately market introduction. The process of gaining regulatory clearance or satisfying its standards is highly precise and apparent. Each country has its own regulatory agency that is tasked with the duty of enforcing laws and regulations as well as providing the essential guidelines for the marketing of pharmaceutical products. This book discusses the process of drug approval and the regulatory obligations that must be fulfilled in line with the United States Food and Drug Administration (USFDA), the European Medicines Agency (EMA), and the Central Drugs Standard Control Organization (CDSCO), India. A significant proportion of researchers either discontinue their work midway through the procedure or have difficulties adhering to the established protocols. Hence, it is unsurprising that a thorough comprehension of regulatory prerequisites is important for any biological researcher involved in drug development. This book aims to encompass all the necessary regulatory standards and offer up-to-date information regarding the mandated regulation of herbal medicines.

The historical background of the establishment of regulatory agencies in India is a subject of considerable academic interest, since it provides insights into the chronological progression of events. The current legislation, known as the Drugs and Cosmetics Act of 1940, together with its later changes, emerged as a result of the efforts of the Professor R. N. Chopra committee. This committee primarily aimed to establish a straightforward quality control system for the herbal chemical "Quinine". The committee's suggestions eventually led to the enactment of this legislation, transforming it into a legally binding framework in India.

There have been instances where the jurisdiction of Ayurveda, Yoga, Naturopathy, Unani, Siddha, and homeopathy (AYUSH), and the Drug Controller Government of India (DCGI), India, have intersected. AYUSH represents and controls the traditional Indian system of medicine, which traces its origins back to ancient times and is primarily rooted in the use of plant-based remedies. Although both processes serve as governmental means to regulate herbal pharmaceuticals, it is evident that each possesses distinct limitations. Consequently, the establishment of a clearly defined transboundary could enhance readers' understanding of this matter.

In addition to its other contributions, the book effectively emphasizes the legal methodology and practical application of herbal medicinal items that have been standardized globally following the Beijing Declaration and the implementation of

AYUSH regulations. The regulation of ayurvedic medications in India is carried out in accordance with the Drugs and Cosmetics Act of 1940, which was implemented in 1982. The aforementioned legislation can be located under chapter IVA of section 33-C, and its parameters are delineated in accordance with the provisions outlined in the first schedule. This particular volume of the book additionally elucidated noteworthy aspects pertaining to the discourse surrounding the functions of ayurvedic medicines and the procedures involved in acquiring regulatory endorsement. Furthermore, the book also emphasizes that drug molecules that are not encompassed within the AYUSH framework but are rather classified as phytopharmaceuticals are subject to the same classification as new drugs. Consequently, the regulatory guidelines governing these substances are analogous to those governing novel pharmaceuticals.

The boundaries between food and herbal medications are also covered in this book volume, along with how their regulatory windows and safety standards differ from one another. The part about the Insecticide Act and herbal substances is very important because it looks into how herbal treatments like neem, karanidin, annonin, rotenone, pyrethrins, and nicotine might work as insecticides. Finally, this particular chapter of the book addresses other mandatory regulatory obligations, including pre-clinical toxicity testing, clinical trials, and stability studies, all of which adhere to the prescribed guidelines.

The chapter addressing the regulatory implications of the approval process in this book will serve as a valuable resource for researchers and students, especially those with expertise in pharmaceuticals, forensic medicine, regulatory affairs, or individuals aspiring to excel in drug research. Furthermore, the contents within this book hold significant potential to captivate a researcher engaged in the field of the herbal drug industry.

Pronobesh Chattopadhyay, Ph.D FNASc, FICS
Danswrang Goyary, Ph.D

Disclaimer

The contents of this book are for information purposes only and are not intended to be legal advice. You understand that this book is not intended as a substitute for consultation. The publisher and the authors strongly recommend that you consult with your legal firm/advocate or regulatory authority before beginning any exercise for approval. The authors are not legal advocates or regulators and represent that they have no expertise to exercise for regulatory clearance. Thus, the information and/or documents contained in this book do not constitute legal advice and should never be used without first consulting with a professional to determine what may be best for your individual needs. Further, this book does not constitute legal advice and no attorney–client relationship is formed, and the publisher and the authors assume no responsibility for errors, inaccuracies, omissions, or any other inconsistencies herein and hereby disclaim any liability to any party for any loss, damage, or disruption caused by errors or omissions, whether such errors or omissions result from negligence or any other cause. The opinions and assertions presented in this book are exclusively those of the authors and may not necessarily reflect the viewpoints of their respective connected or affiliated institutions or the publisher, editors, and reviewers involved. The publication neither warrants nor supports any assertion or information offered here.

Pronobesh Chattopadhyay, Ph. D. , FNASc, FICS
Danswrang Goyary, Ph. D.

1 Process of Drug Discovery

1.1 INTRODUCTION

The discovery of new medicine is a result of typical systematic scientific methods whereby interdisciplinary knowledge and attempts are needed. It new medicine is a lengthy process. A small number of molecules—say, about 10 of 10,000 ingredients/molecules—are usually recognized as prospective drugs after screening several times and will make it to the human trials. Safety aspects are the most important to ensure that the new molecules have no potential toxicities. If any molecules are reported to have serious adverse reactions, they are removed without any further examination. Eventually, only about 1 in 10,000 molecules may be recognized as prospective medication after clinical trials and found to be valuable, while the remaining 9,999 are refused for one reason or another.

It is a Herculean task and a real challenge to discover and develop the new medicines rather than manufacturing them. Application of computer-aided drug design, modern software, identification of pharmacophores, and good apothecaries can carefully design a new molecule in a matter of a few months, a compound that has been found from among millions of others and carefully designed through decades of analysis to prove that it is secure for individual use and efficient as a cure for illness.

The discoverer enjoys the right to discovery and patents provide up to 20 years for exclusive market rights and all privileges of a new medication, during this period others are banned from the marketing of the same medication. New drugs are trademarked to the discoverer as soon as they are invented, and the discoverer usually has less than 10 years of exclusive marketing privileges remaining by the time regulatory authorities accept the medication for promotion. Patents do not prevent others from creating a similar medication with minimal variations in substance framework. However, a drug with even the least difference in substance framework will have to turn out to be safe and efficient through a whole series of experiments.

1.2 DRUG DISCOVERY PROCESS

1.2.1 Screening Large Compounds and Natural Compounds

The desired active molecules can be derived from plants, animal products, or microorganisms. Natural products, having large-scale structural diversity and more than synthetic compounds, have been the major resources of drug molecules and most of the bioactive molecules originate from natural sources only. The advancement of synthetic chemistry and molecular biology has appeared as new technology to initiate receptor-based targets of active compounds as probes to characterize proteome

DOI: 10.1201/9781003398431-1

functions. Computer-aided drug design shortens the tedious steps for synthesis and in this technique the desired molecules (lead molecules) are generated and predicted using computers for possible receptor binding and pharmacological response.

1.2.2 Modifying the Chemical Structures of Existing Molecules

Combinatorial chemistry consists of systemic methods for the possibility of preparing a large number (tens to thousands or even millions) of existing compounds in a single process. Also, combinatorial chemistry can be used for modifying small molecules to increase bioactivity, has been shown to enhance several features, including but not limited to greater absorption, safer use, extended length of action, enhanced effectiveness, particularly in the context of peptides.

Also, structural modification in combinatorial chemistry alters the action of a compound. After identifying the lead compound, optimize the desirable characteristics of the compound. Combinatorial biosynthesis serves as an essential link between chemical compounds and the evaluation of potential therapeutic candidates through the process of modeling and constructing libraries. The linkage between different biological activities is established by the incorporation of chemical structural information.

1.2.3 Studying Disease Pathogenesis

1. Disease pathogenesis is closely associated with the alternation of biochemical pathways and receptor unresponsiveness or overexpression. Identification and targeting of the pathways are closely associated with drug discovery.
2. After identifying and validating the target, the drug is synthesized for the target. The novelty of the drug discovery is that the synthesized chemical compounds will have biological activities and be effective in targeting particular stages of a disease process.

1.2.4 Modern Techniques of Drug Discovery

1.2.4.1 Molecular Modeling

The approach of by using software modeling and graphics tools to simulate the affinities of various receptors and create 3D architectures of target molecules or targets provides a new dimension of the drug discovery.

1.2.4.2 Medical Genetics

Pathogenesis, severity, or resistance of the disease are unknown to individuals. The answer to the mystery is the using genomics and proteomics approaches toward drug discovery. Genetic linkage studies are useful to understand the link genes with particular diseases.

1.2.4.3 Robotic High-Throughput Screening (RHTS/HTS)

High-throughput screening (HTS) is the rapid technique of drug discovery by using biology, chemistry, and pharmacology by using software, liquid handling devices, and sensitive detectors. Pharmacological tests, bioactivity, and preliminary toxicity of the lead molecules can be tested quickly and identified accordingly. Also, HTS helps to identify the active compounds, antibodies, or genes which are involved in the pathogenesis of the disease and understanding the particular biochemical process or interaction in the disease conditions. HTS is the process of assaying a large number of potential lead compounds, and more than 20,000 compounds per week are added to the library.

1.2.5 Clinical Steps of Drug Discovery

Laboratory animal studies are an integral part of the drug discovery process, and they often serve as the initial steps in assessing the safety and efficacy of investigational products. Small animals, including albino rodents such as mice and rats, are the most popular method for confirming a product's safety for use in humans.

The use of animals has reduced over the decades as new alternatives to animal studies have become available. Animal experimentation and examination can be removed only for a community of non-clinical research; animal toxicology assessments are still considered essential to drug growth and are needed by government authorities before they will allow for a human trial. The great majorities of lead molecules fail *in vivo* or animal study and lead to attrition in the direction. Sometimes development initiatives have to be discontinued and discovery work re-initiated because all potential applicants failed non-clinical examination.

After successful examination of pre-clinical study, the drug is prepared for clinical study. An essential document for regulatory approval in the process of obtaining authorization to perform human trials is the Investigational New Drug Application (INDA) of formulation including tablets, capsules, or injection. After evaluation in the animal model, only safe and potential compounds will enter into clinical examination or trial. Government and regulating authorities thoroughly examine the results of non-clinical study and except them for further human trial. The risk-to-benefit ratio also is calculated, and only those molecules for which authorities believe the potential advantages to the patients will be greater than potential side effects are considered.

A few healthy human volunteers (30–50) are recruited in a Phase I trial by applying a small dose of the test compounds in a specific, in the presence of medical professionals who have expertise in the first-in-human research. The benefits of the undertaken research and all its risks and advantages are briefed to volunteers who are compensated with a participation fee if they decide to join. Furthermore, recently, DCGI has given a provision to patients with insurance if any accident occurs during the trial. Dosing of the test drug is slowly increased over several days until the frequency of minimal side effects gets to the upper end of the acceptable range, or the full dose range is achieved. The nature of any adverse reactions and the drug focus in the human body is revealed.

Subsequently, after confirming that investigational molecules are free from side effects, they are introduced in Phase II studies, a step allowed only when impartial expert and government regulators continue to determine that the potential benefits for patients outweigh the possible risks of adverse effects. Phase II studies are conducted on several hundred volunteers who are actually afflicted by the illness for

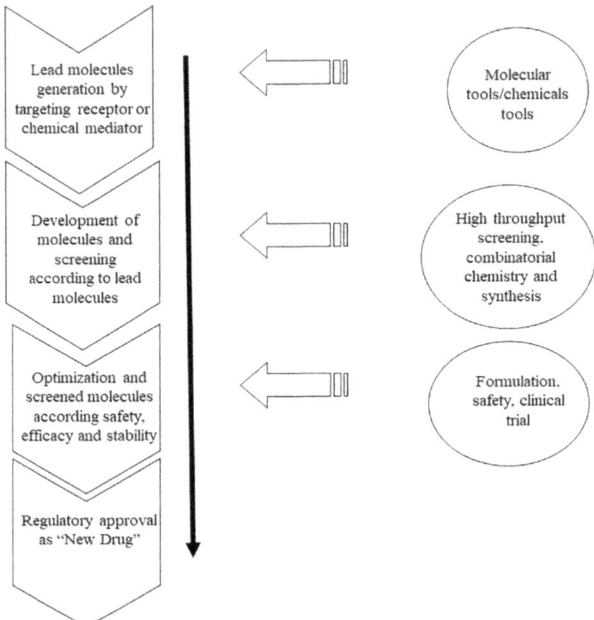

FIGURE 1.1 The drug discovery process and regulatory clearance.

which the investigational substance is designed. The research and the investigational medication are thoroughly explained to the subjects, outlining both potential benefits and possible adverse reactions. Those individuals who wish to participate in the research are then enrolled. Patients receive free treatment and all blood and other biochemical assessments (including urine and stool samples, etc.) and sponsoring company is responsible for the financial expenses of all investigations of the volunteers. In this phase, the volunteer will neither be paid any fees nor be compensated for contributing to the research. The informed consent papers and patient employment procedures are analyzed by government authorities and the ethics committee of the hospitals in advance.

Phase II studies highlight the effectiveness of medicine and in determining the therapeutic effective dose at which it works best, whereas Phase I confirms the safety of a drug. After successful Phase II studies, if the drug proves to have the fewest side effects and is economically viable and beneficial for therapeutic application, then the drug enters Phase III trials.

Phase III, often known as full development, is the last stage of drug development. The investigational drug is provided to a wide range of individuals, including children and the elderly, those with varying degrees of disease magnitude, those taking other medications for other diseases, those who must take the treatment for an extended period of time, and so on. Numerous venues are used to recruit patients for studies, simultaneously in many countries across continents, which are known as a multicentric trial. The new medicine is compared with older drugs to confirm the effectiveness and is known as bioequivalence.

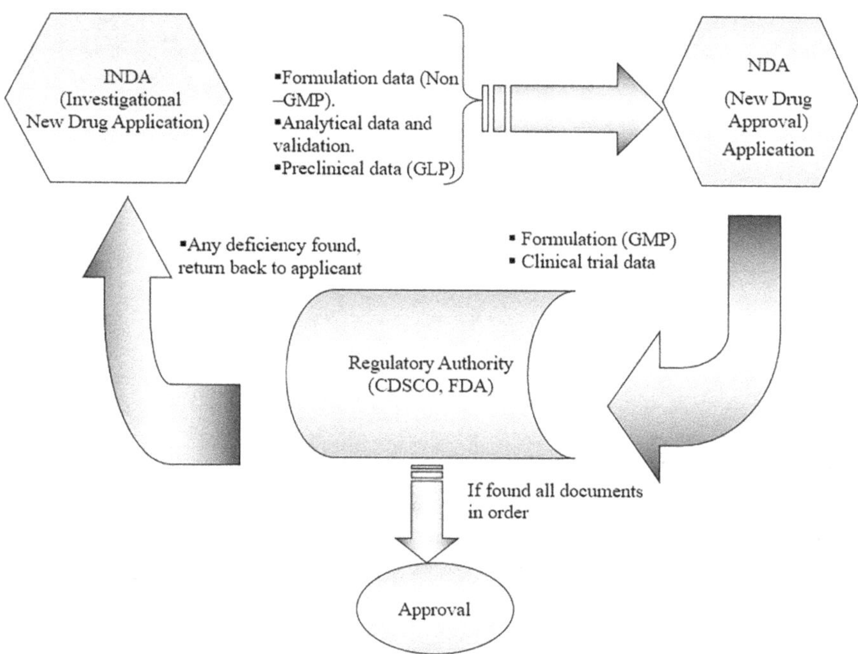

FIGURE 1.2 Outline of the investigational new drug (IND) application process.

If it turns out that the drug is just marginally better than prior medications, development of the drug may have to be discontinued.

Phase II clinical trials outlined the potency and efficacy of the medication, identifying the optimized therapeutic range at which molecules work best, whereas Phase I verified the safety of the medication. After completion of the Phase II trial and getting the full information of the molecule, including its effectiveness and dose margin, the molecules enter into Phase III trials or full growth. In this phase, a wide range of age group volunteers are recruited comprising children and adult populations. Also, three other investigational points are taken into consideration: (i) the effectiveness of the drugs in different progressions of the disease; (ii) drug interaction of the investigational drugs with other drugs or medications used by the volunteers for other illnesses that are needed to be taken for a long period, and so on; and (iii) food and drug interactions which are usually noticed in the therapeutic efficiency of the investigational molecules with the food taken. Phase III trials are performed by wide advertisement to recruit maximum volunteers. The investigational new medicine may get approval if it is shown to be only as good as other established medication having less expense. In India, most of the medicines are bioequivalent medicine, and recently, new molecules are entering the market gradually.

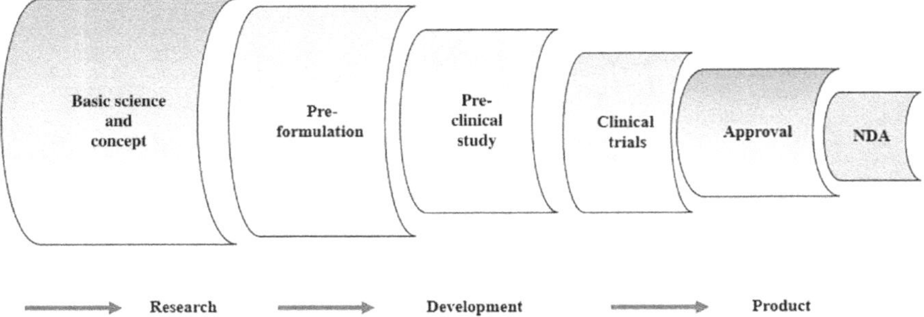

FIGURE 1.3 Interdisciplinary approach to the drug discovery process.

1.2.6 Post Approval Studies

Post-marketing surveillance is an important aspect of marketing drugs. Regulatory authorities keep a tight watch on the newly approved medicines. Physicians are needed to evaluate and report any unexpected complications or alleged hazards to wellness with the new medication—known as adverse drug reaction (ADR)—as soon as possible to the regulatory authorities and the drug manufacturer involved. Any individuals who begin a new medication face a new risk of health hazards which continue to persist throughout the world. Regulators observe carefully and sometimes alerts and safety measures must be added to the product label. In some cases, a product may have to be removed from the market due to complicacy in medication; this may happen when the new drug that may not have been engaged in Phase I–III trials in new communities (e.g., children, older, pregnant women). Regulatory authority–enforced drug manufacturers perform Phase IV trials as a mandatory condition for approval. Companies often perform post-marketing trials without regulatory controls. A post-marketing trial analyzes a promoted medication in new age categories or types of patients. Unrevealed or unknown negative impacts or related risk factors are sometimes focused on in this phase. The main purpose of drug development is to ensure the safety and therapeutical effectiveness of the marketed drug. Showing adverse effects in Phase IV trials may result in the withdrawal or restriction of a drug (e.g. troglitazone, celecoxib, etc.). After evaluating the results of all the trials in great detail, including visiting the study sites and the cross-questioning of researchers and investigators by the government authorities and regulatory agencies, marketing authorization is given. The investigational medication is then "launched" and becomes a medication available at the pharmacy or drugstore. New medicines are indeed expensive. Most of the time, the pharmaceutical company increases the price of new medicine for the invested price in drug discovery, promotes the benefits of the new therapeutic invention, and provides profits to investors of the organization. Patent or exclusive right of marketing expires in 8–10 years; most pharmaceutical companies are involved in manufacturing the same molecules, and in this instance, the medicine sells at a low price. People are often not able to manage new medications, and in most nations, the government subsidizes and provides them at a low price to patients.

1.3 CLINICAL TRIALS REGISTRY OF INDIA (CTRI)

The Clinical Trials Registry of India (CTRI) is an official platform and catalog for registering clinical trials conducted in India. Most of the trials are conducted for the generation of information related to the safety and effectiveness of health measures, such as medications, diagnostics, equipment, and therapeutic procedures. The aim of the CTRI is to provide transparency and make access to clinical trials available to the public for assessment of a new drug application (NDA). The data and the progress of the clinical trials are often searchable by disease/indication, drug, location, etc. In India, a large number of contract research organizations (CRO) are in operation, and the maximum hospitals or foundations are involved directly or indirectly in sponsoring the new drug, diagnosis kit, or medical device. The CTRI was established in October 2005 and registration became mandatory for any clinical trial from 15 June 2009. The CTRI is funded jointly by the Department of Science and Technology (DST), the World Health Organization (WHO), and the Indian Council of Medical Research (ICMR) of New Delhi. The national pharmacovigilance program coordinated by Central Drugs Standard Control Organization (CDSCO), under the Directorate General of Health Services, the Ministry of Health and Family Welfare of the government of India is a considerable step for providing the safeguard.

1.4 INVESTIGATIONAL NEW DRUG APPLICATION (INDA)

After completion of a pre-clinical trial, there is a need to file an INDA with the United States Food and Drug Administration (USFDA) to begin testing of the drug in humans. An INDA is mostly technical terms whereby most manufacturers or sponsors manufacture the medicine and the means by which usually the producers or potential marketers acquire a legal position to call their new investigational material as new medication. An INDA is not an application of marketing acceptance. Usually, an INDA contains product details including the composition, manufacturer, stability, and controls for quality control, and the details are evaluated to ensure that the organization continually produces and supplies consistent batches of the medication. Investigator information is crucial for the evaluation of study protocols, including to assess the risk associated with exposure to unnecessary medicine or trials. Clinical trial protocol mentions the detailed methods for suggesting investigation to evaluate whether the initial-phase tests will reveal topics to needless threats. Also, information of the clinical investigators (generally physicians) is necessary to any clinical trial to judge if adequate responsibilities of the clinical trials were met. It is also desirable in the clinical trial that the associated possible risks are mentioned, and finally, the responsibility comes to the institutional review board (IRB) to evaluate for acceptance as per clinical trial rules offered by regulatory authorities.

1.5 NEW DRUG APPLICATION (NDA)

New medicine can be regarded as "new" because of its structure, or use, its dose. It can easily be recognized that a medication containing its component with a new substance enterprise would be regarded as a new drug. Usually, one drug can be designated as new because of the structure of its non-active components, the proportion of components, efficient or non-active, or the combination of components, efficient or non-active. Also, a drug can be designated as new because of the use, route of administration, and dosage forms. After successful demonstration of the safety and effectiveness of the molecule in a Phase III trial, the company usually files with the USFDA for approval of a NDA. The data required for filling out NDAs encompass both nonclinical (animal) and clinical (human) test data, which include analyses, drug details, and documentation of manufacturing techniques.

An NDA filing typically runs around 100,000 pages or more. Usually, the USFDA takes six months to review, and in most cases the period between the first submission of an NDA and final USFDA approval exceeds that limit.

An NDA is a regulatory mechanism established by the USFDA that enables drug sponsors to formally submit and seek approval for a new pharmaceutical product to be marketed and sold in the United States. Filing of NDA is an art and particularly the inspector investigators of USFDA/DCGI look into the issues regarding the safety, effectiveness, and claims of the manufacturer. One should be very particular about the following when filing an NDA:

1. Safety and effectiveness for its proposed use(s), and whether the benefits of the drug outweigh its risks.
2. Proposed labeling and active ingredient disclosure.
3. The manufacturing process and the process variables to control the quality, including identity, strength, quality, and purity.
4. Data of pre-clinical (animal) studies and all clinical (human) studies. The information in summary form or abstract form must be provided to ensure that the full case review of each person who obtained the medication is required only in restricted circumstances. Reports formerly presented to the USFDA in the IND or regular reviews must be engaged by reference in the NDA. The details included therein must be adequate to rationalize the claims made in the suggested marking for the molecules as to performance, therapeutic efficacy, dosage, and foreign information that can be used. The label was decided by an exchange between the sponsor and the USFDA. Supplementary new drug application is also common phenomena to USFDA filing, which is concerned, particularly any essential modification in the production, control, product packaging, or other actual properties of the drug, or any modification of its marking that may affect therapeutical effectiveness and performance compared to either the drug itself or how it is used.

1.6 OVER-THE-COUNTER (OTC) DRUGS

Regulatory requirements of an NDA for prescription and over-the-counter medication are similar. It is incredibly rare, however, that any new substance enterprise would be accepted by the USFDA for over-the-counter sales for aspects of lack of adequate information to support the safety in the use of a totally "new" medication. The label contains adequate direction for use and therapeutic intervention. Almost all OTC filling often involves the provision of extra filling, mostly based on the premise that the drug can be properly utilized without the need for direct medical supervision.

1.7 ABBREVIATED NEW DRUG APPLICATION (ANDA)

Filing of an ANDA is somewhat simpler than others and requires fewer documents or data because formerly obtained information about the particular molecule has been registered with the USFDA. The purpose of the USFDA is to reduce replication of attempts in drug application about which some of the required details are already available while guaranteeing that the new items will be comparative to an approved drug. Generally, the inclusion of animal and clinical data is not mandatory for establishing the safety as well as efficacy of an ANDA. Most of the ANDAs are very close to generic products and USFDA does not encourage reverifying the efficacy and safety-related issues; the foremost requirements are labeling, specifications, and the manufacturing process, other than generic active ingredient bioequivalence of the new product relative to an NDA-approved product. Thus, ANDA filing is an approach to ensure that with active ingredients—which have been available in the industry for many decades and in many items—are recognized general medication, it is needless to reverify their effectiveness.

Compared with the NDA filing which needs a large amount of data including safety aspects, clinical effectiveness trial, stability data, etc., an ANDA only needs few data; the active ingredient's chemistry, source of supply, description of the dosage form to be marketed, and a brief statement of the manufacturing place are required for an ANDA filing. Also, the quality controls to assurances that the drug will comply with reference active ingredients and appropriate specifications. Further, the requirement of ANDA filing is mandatory for compliance with a good manufacturing process (Schedule M of Drugs and Cosmetic Act of 1940) including methods, details, packaging, and bioavailability.

The characteristic of the drug, method of use, method of manufacture and packaging, stability, the extent of use, safety and potency, and history is a considerable approach for the filing of an ANDA, and deviation of these lead to not register the drug in the ANDA category.

It is a fact that most drugs become cheaper for ANDAs and reduce drug development costs to explore unnecessary clinical and safety trials. It is more correct that most Indian companies are exploring their potential in an ANDA filing rather than an NDA filing. Partially, it is true that in the Indian context, the manufacturer is attempting to promote replication of drugs and to create drugs which are not protected by patents and are easily obtainable to the consumer in a competitive industry.

1.8 GENERIC DRUG PRODUCT

The genesis of generic medicine had stated with an amendment of the Food, Drug, and Cosmetic Act 1984, entitled "Drug Price Competition and Patent Term Restoration Act of 1984". Further, the entire theme of generic medicine was that they are produced with a minimum amount of clinical effort but with the establishment of the physical and chemical equivalency of the approved drug, and this is amended as the Waxman-Hatch Amendment. It is true that such a noble and basic thought has changed the entire process of ANDA filing.

In India, the Drugs (Prices Control) Order of 1966 and the Drugs (Prices Control) Order of 1970 were validated under the Essential Commodities Act 1955 considering drugs to be essential commodities. Thereafter, issued national "Drug Policies" from time to time ensure effective control measures concerning the cost of the required medicines and subsequently bulk medicines, as well as the accessibility of medications, all the while considering the development of an innovative, cost effective, and capacity building domestic industry.

It is true that Indian drug policies are highly effective and in the context of the liberalization of the economy and the allowing of foreign investment in the country, drugs are becoming cheaper. Most generic medicine comes from the new drug approval process when the patent is expired and nevertheless, the generic medicine replaces the marketed expensive drugs and sometimes substitutes the expensive drugs. Bioequivalence may differ from the other branded or established drug. This act was also folded into the patent restoration act, which allows an extended chance to be given to drug manufacturers for new medication ingredients or excipients. Not only does this act restore the generic drug, but it may differ from the branded or established drug in physical appearance (i.e., size, color, shape). Generic brands are economic because the investment costs of drug development are not required and it is evident that this act encourages many firms to manufacture generic drugs, but by regulation, the manufacturers who filed for the ANDA and met the specific norms of filing are allowed to market the generic drug. A generic drug should be the bioequivalent of the branded product, and similar therapeutic effectiveness and clinical responses are expected. A generic drug is comparable to branded or the dose form, strengths, delivery route, quality, and efficacy characteristics of existing medications, and clinical applications. Generics have the same risks and benefits as their brand-name counterparts because the active pharmaceutical ingredient (API) is similar. Generic drugs are manufactured in the same dosage form as established available drugs in the market (e.g. liquids, solid, aerosol, or injectables). Therefore, supporting bioequivalence, generic drug trials are limited to a small number of healthy human volunteers. The manufacture should submit the ANDA to the DCGI or/ and USFDA for approval to manufacture the generic medicine.

1.9 FIXED-DOSE COMBINATIONS (FDCs)

More than one active API are present in the formulation known as a fixed-dose combination (FDC) drug. FDC drug as a typical and approval process is quite different. According to the category, the approval process is also different.

1. In an FDC formulation, any active ingredients are treated as a new drug and in such cases similar data is required for an NDA including clinical trials for approval of the FDC drug.
2. Drugs not marketed as FDC but the individual API is already in use concurrently, and for the said claim, the approval marketing permission may be granted based on chemical and pharmaceutical data. The stability data of the FDC dosage form is essential for approval.
3. Clinical trials may be required in the case of other FDCs. Based on DCGI or USFDA pharmacological, toxicological, and clinical data on the individual

ingredients, the clinical trial permission is granted. The rationale for combining and the proposed ratio of the API should be mentioned. Additionally, individual ingredients, as well as their combination in the proposed ratio pre-clinical data of acute toxicity data (LD_{50}) and pharmacological safety data, are required.

4. With any alternation of marketed FDCs, including the changing of the ratio of active ingredients, for making a new therapeutic claim, in order to receive marketing authorization, it is necessary to present the reason for the fixed-dose combination (FDC) along with published reports. Depending upon the submitted data, FDC may be approved according to the claim.
5. Development or approval of a Fixed-Dose Combination (FDC) of Active Pharmaceutical Ingredients (APIs) for a specific indication. The FDC involves APIs from the same class, and these APIs have been used widely for a particular indication over the years. The intention is not to make any claims beyond expediency in a particular category. This suggests that the FDC is being considered for its practical advantages in a specific therapeutic area. The API exhibits stability, which is a crucial factor in drug development. Stability ensures that the drug maintains its quality and efficacy over time. The components of the FDC demonstrate a small chance of substantial pharmacodynamic interactions. This is important to minimize the risk of unintended effects when combining multiple drugs. There are no indications of significant alterations in bioavailability. Maintaining consistent bioavailability is critical for ensuring that the drug achieves the desired therapeutic effects. The proposal suggests that, given the stability of the API, limited pharmacodynamic interactions, and no significant alterations in bioavailability, the FDC can be approved without the need for additional animal or human data.

1.10 USFDA

1.10.1 Drug Discovery Process: Code of Federal Regulations (CFR)

The Code of Federal Regulations (CFR) is the permanent rules and regulations published in the Federal Register by the different departments or agencies of the federal government of the United States. The CFR is categorized into 50 titles; among them 21 are concerned with drugs and related issues.

In general, 21 CFR construes the federal food, drug, and cosmetic act and its antecedent laws. Section 2 of the CFR contains the majority of the regulations pertaining to food and medications, and the USFDA's part of the CFR interprets the Federal Food, Drug, and Cosmetic Act and related statutes. The CFR title 21 is organized into three distinct chapters.

Chapter I: Food and drug administration
Chapter II: Drug enforcement administration
Chapter III: Office of national drug control policy

The following regulations are important for INDA or NDA application.

1. 21 CFR Part 312: Investigational New Drug Application (INDA)
2. 21 CFR Part 314: INDA and NDA applications for USFDA approval to market a new drug
3. 21 CFR Part 316: Orphan drugs
4. 21 CFR Part 58: Good lab practice for non-clinical laboratory (animal) studies
5. 21 CFR Part 50: Protection of human subjects
6. 21 CFR Part 56: Institutional review boards
7. 21 CFR Part 201: Drug labeling
8. 21 CFR Part 54: Final disclosure by clinical investigators

BIBLIOGRAPHY

1. McInnes C. Virtual screening strategies in drug discovery. *Current Opinion in Chemical Biology* 2007; 11: 494–502.
2. Reich JW. *Remington: The Science and Practice of Pharmacy* 1995; (19th Edn): 1: 52–59. Lippincott Williams & Wilkins.
3. Jaime ND, William AR. *Wilson and Gisvold's Textbook of Organic Medicinal and Pharmaceutical Chemistry*. J. B. Lippincott Company 1991; (9th Edn): 1–2.
4. William RP, Raymond DM. *Drug Regulatory Affairs. The Theory and Practice of Industrial Pharmacy by Leon Lachman, Liberman and Kanig*. Verghese Publishing House 1987; (3rd Edn): 856–882.
5. Williams PD. The role of pharmacological profiling in safety assessment. *Regulatory Toxicology and Pharmacology* 1990; 12: 238–252.
6. Singh SS. Preclinical pharmacokinetics: An approach towards safer and efficacious drugs. *Current Drug Metabolism* 2006; 7: 165–182.
7. Murphy MP. Current pharmacogenomic approaches to clinical drug development. *Pharmacogenomics* 2000; 1(2): 115–123.
8. Christoph MH. Synthetic library design. *Drug Discovery Today* 2006; 11: 763–767.
9. Kuhlmann J. Alternative strategies in drug development: Clinical pharmacological aspects. *International Journal of Clinical Pharmacology and Therapeutics* 1999; 37: 575–583.
10. The Drug Development and Approval Process. (www.mhsource.com/resource/process.html).
11. *From Test Tube to Patient*. A Special Report from FDA Consumer Magazine. (www.fda.gov/fdac/special/testtubetopatient/studies.html).
12. Ross T. An overview of the drug development process. *The Physician Executive* 2006; 48–52.
13. Michael RS, Tornell J. The use of transgenic systems in pharmaceutical research. *Briefings in Functional Genomics and Proteomics* 2002; 2: 119–130.
14. Center For Drug Evaluation and Research (CDER). Frequently Asked Questions on Drug Development and Investigational New Drug Applications.
15. Center For Drug Evaluation and Research (CDER). Generic Drugs: Questions and Answers.
16. Center For Drug Evaluation and Research (CDER). Think It's Easy Becoming a Generic Drug in America? Think Again.
17. Allergan. Drug Development Process.

18. Drugs and Cosmetic Rules. 2nd Amendment, Ministry of Health and Family Welfare (Department of Health), notification, 20 January 2005.
19. Julka PK. Clinical trials in India, dilemmas for developing countries. *Issues in Clinical Research* 2007; 69–71.
20. ICH topic E6, Guideline for good clinical practice, the European agency for the evaluation of medicinal products, step 5, consolidated guideline 1.5.96.
21. Guarino RA. *New Drug Approval Process*. Marcel Dekker Incorporation. (2nd Edn): 39, 2005.
22. Registration of Clinical Trials. *The Pharma Review* 2007; 6: 29. (Kongposh Publications, ISSN No. 0973–399X, 52–54).

2 History of Drug Regulatory Authorities

2.1 INTRODUCTION

Clinical trials are precise and scientific studies for drug development for the assessment of safety (adverse effects or reaction) and efficacy data to be collected for health status and clinical trials done with drugs, diagnostics, devices, analysis, and protocols. After the success of the pre-clinical study, the clinical trials are to be done and approval is granted in the country where the trial is taking place. Outcomes of the clinical trial are properly acknowledged, along with the disease pattern and the natural history of untreated disease.

The first reference to a clinical trial was reported by King Nebuchadnezzar II (605–562 BCE) in the Bible. He instructed a group of young individuals to follow a meticulously regimented diet, consisting solely of meat and alcohol, for the duration of three years. This particular group of children was then subjected to a comparative analysis with three more youths who were restricted to the royal edict, choosing instead to consume a diet comprised exclusively of pulses like beans, peas, and legumes, and water.

After just 10 days, the monarch made an observation that individuals who consumed a diet consisting of pulses and water exhibited superior physical fitness when compared to those who had followed a diet comprising meat and alcohol. Consequently, the researcher concluded the experiment and transitioned the participants who had previously been provided with meat and alcohol to a diet primarily consisting of pulses.

Al-Quanun fi al-Tibb or *The Canon of Medicine* by Ibn Sina (Avicenna), a Persian scientist, demonstrated the thorough compilation of all known medical information and included the Arabic medical knowledge with personal experience in this book. Also, he acknowledged the knowledge imparted in the medical science of Hippocrates, Galen, Dioscorides, and others. He highlighted the testing of new drugs on animals and humans before clinical application and advised following the seven main features of any drug testing. He emphasized that the drug should be pure and should be used on a "simple" disease. Furthermore, the drug should be evaluated on at least two different types of diseases, and the drug's efficacy should match the severity of the diseases. He also pointed out that the at-the-time determination of efficacy of the drug's effects on natural healing should be ruled out and uniform observation should be obtained in each trial. Perhaps he was the first scientist who pointed out that in clinical trial and drug should be tested in animals first followed by humans, and simultaneously that the trial should not be done on animals and humans. This book was considered as an important standard by regulatory authorities for a century in Europe, as well as the Middle East.

DOI: 10.1201/9781003398431-2

James Lind conceptualized the modern clinical trial for the first time. In his trial in 1747, he concluded that citrus fruits could be helpful to treat scurvy in patients onboard a British naval vessel. He compared the effects of the other food supplementation such as cider, borsch vitriol, vinegar, seawater, nutmeg, and citrus baked apples (oranges and lemons). After 50 years, the British Navy implemented lemon juice as a requisite part of the seaman's diet.

The concept of placebo in the clinical trial is very important. Austin Flint in 1863 first conceptualized placebo therapy. In a comparison study of the treatment of rheumatic fever in 13 patients, conclusions arose which compared the findings to those reported after receiving active treatment. Throughout this study, also attention was posted to the expected possibility of active treatment and effectiveness in preventing complications including pericarditis, endocarditis, pneumonia, etc.

The concept of randomization is another important aspect of clinical trial. R. A. Fisher in 1923 conceptualized randomization in a study setting with his experiment of crop variation.

At the time of World War II, horrific human experiments were done in the name of science. The Nuremberg Code was established for the global response and is the ethical guideline accepted throughout the world for the prevention of atrocities in the name of science. The Nuremberg Code was created in 1947 in the purview of stopping war crimes involving human patients as "volunteers". An effort of Dr. Leo Alexander in 1947 raised issues to the right of war crimes and persuasion in medical research. Later, issues pointed to some humanitarian facts including informed consent, benevolence toward the confidentiality of the experiment, and proper design of the experiment.

The declaration of Helsinki was another landmark point of the clinical trial history and was formed by the World Medical Association in 1964—a total of eight revisions have already been made. This is another safeguard to volunteers, and the declaration was a set of ethical principles of human experimentation. The declaration promotes the formation of institutional review boards (IRBs) or Institutional Ethical Committees (IECs) to protect the interests of patients or volunteers, and rational recruitment of the volunteers. Moreover, this declaration is an extension of the Nuremberg Code by applying the doctrines specifically to clinical research. The declaration has great importance and imposes a regulatory mechanism to the medical community to regulate drug trials. Looking to the international regulations, most countries made their regulations of clinical trials to address ethical issues and prevent exploitation of human subjects. The most influential institutions are the United States Food and Drug Administration (USFDA), the European Medicines Agency (EMEA), Japan's Pharmaceuticals and Medical Devices Agency (PMDA), and India's Central Drugs Standard Control Organization.

1. Egypt 1550 BC—Egypt was very rich in chemical formulary and medicine. The kings of Egypt usually used medicine for the healing of the sick, and they documented the prescriptions. Some written documents of related medicine and science are seen in the grave of the kings (Egyptian pyramids).
2. 600 BC—The Book of Daniel in the Bible describes the comparative trial of food consumption, a vegetarian diet or a diet rich with meats and wine,

perhaps the first concept of the clinical trial. A ten-year study revealed that the vegetarian diet was accepted as the healthiest.

3. 1700s—In 1754, surgeon James Lind's experiment on the effect of citrus fruit in treating scurvy aboard the *HMS Salisbury* later introduced citrus food for British sailors. Another landmark breakthrough the concept arose in the vaccination development by Edward Jenner's experiments in 1976, a landmark trial which concluded that inoculation could prevent smallpox. Jenner's trial was indeed the first concept toward mass vaccinations programs of the other conquered epidemics including typhoid, polio, measles, and yellow fever.
4. 1900s—The modern clinical trial was conceptualized from this era and Sir Austin Bradford Hill, British epidemiologist, in 1946 categorized the populations of a clinical trial into experimental and control groups at random so that the bias of the experiment is nullified. The results and outcome of the clinical trial were clearer, and differences were seen in the health of the two groups. The guidelines of the clinical trial were established, and different regulations came into force. Thousands of new medicines and therapeutics were approved.

TABLE 2.1
Chronology of the Breakthrough of the Drug Discovery and Regulatory Advancement of the Drug Discovery

Chronology	Events
600 BCE	First clinical trial mentioned in the Bible: King Nebuchadnezzar II and Daniel compared the diet of vegetarians and non-vegetarians.
10th century	*Canon of Medicine* written by Ibn Sina.
1557	First hospital in Japan established by Europeans.
1747	Conceptualization of with untreated control group by Carl Linnaeus (Carl von Linné), a Swedish botanist, physician, and zoologist, 1753.
1799	Conceptualization of sham procedure by Haygarth, 1800.
1813	Regulation of US Vaccine Act.
1820	Origin of US Pharmacopoeia.
1862	US Division of Chemistry (predecessor of the USFDA) formed.
1863	Conceptualization of placebo treatment by Sutton, 1865.
1874	Regulation of drugs in Japan initiated by the Meiji government.
1883	Concept of randomization introduced by Charles Peirce and Joseph Jastrow.
1902	US Biologics Act for government pre-market approval for drugs.
1906	Enforcement of US Pure Food and Drugs Act in the judiciary system.
1923	Concept of randomization by Fisher and MacKenzie.
1931	Medical Research Council formed by the British government to control the clinical trial.
1937	National Cancer Institute (National Institutes of Health, 1981b) is formed and promoted drug research.
1943	First regulatory law comes into force in Japan.
1946	Nuremberg Code formed for human experimentation.

TABLE 2.1 (*Continued*)
Chronology of the Breakthrough of the Drug Discovery and Regulatory Advancement of the Drug Discovery

Chronology	Events
1948	First time "proper" experiment randomization techniques implanted by the British.
1954	Pesticide amendments introduced in the United States.
1962	Food, Drug and Cosmetic Act amended (US Congress).
1957–1961	Thalidomide disaster.
1966	US Public Health Service regulations on institutional review boards for research involving humans subjects.
1980	Controlled clinical trials.

2.2 HISTORY OF THE UNITED STATES FOOD AND DRUG ADMINISTRATION (USFDA)

In the 19th century little dominance or harmonized controlled measures existed over food and medications. The individual states were concerned about food and medicines, and varied from state to state and not were uniform. The admixture and misbranding of foods and drugs were rampant. Individual pharmacies also manufactured drugs without measures to control quality or effectiveness. The aboriginal federal law became the Vaccine Act of 1813 with the aim to protect consumers' rights, including safety and quality, and extension of the law established a national source for an uncontaminated smallpox vaccine. However, the Vaccine Act was repealed after nine years because of a baleful blow and public scandal involving contaminated vaccines.

President Lincoln in 1862 formed the Division of Chemistry, the antecedent of the USFDA, and in 1867 the new Department of Agriculture formed to police the bribery of agronomical commodities. Harvey Washington Wiley in his role as arch chemist proved the analytic role of the Division of Chemistry in 1883. He was active in the framed of the Biological Act of 1902, an acknowledgment of the deaths of several accouchements acquired by spurious smallpox vaccines and diphtheria antitoxins. The Act was implanted to the federal government for pre-market entry or approval for every biological and approval for the production. He wrote a famous book, *Foods and Food Adulterants*, published in ten parts and appearing from 1887 through 1902. The effectiveness of varying food additives administered to healthy volunteers is well documented. His contribution to organizing the US drug regulatory authority is unquestionable. He conceptualized and unified state chemists, food and drug inspectors, the General Federation of Women's Clubs, and national associations of physicians and pharmacists. He was a pioneer of enforcement of the Pure Food and Drugs Act, or the Wiley Act, which implement by President Theodore Roosevelt in 1906. The law was so effective that immediately the US Pharmacopoeia (USP, originated in 1820) and the National Formulary were recognized. Further, the law was strengthened by inducing the prosecution of misbranding or the spurious formulation of drugs and false advertising. The action pertained only to the drug labels, and thus the USP and NF were recognized as official standards for the strength, quality, and purity of drugs. Formulations failing to comply with the USP specifications are known as adulterated drugs.

2.3 CREATION OF THE EUROPEAN MEDICINES AGENCY (EMA)

The European Union (EU) was conceptualized in 1951 when six countries (Belgium, France, West Germany, Italy, Luxembourg, and the Netherlands) created the European Coal and Steel Community for the common market. Through the Rome Treaties in 1957, the European Economic Community (EEC), the predecessor to the EU, joined hands with the common market to develop economic growth. In 1973, the United Kingdom, Ireland, and Denmark abutted the EU; Greece abutted in 1981; Portugal and Spain in 1986; and Austria, Finland, and Sweden abutted in 1995. In 2004, the Czech Republic, Hungary, Estonia, Lithuania, Latvia, Malta, Poland, the Republic of Cyprus, the Slovak Republic, and Slovenia abutted the EU; and, in 2007, Bulgaria and Romania abutted for a total of 27 countries or affiliate states, although the EMA was not formalized until 1995; perhaps the thalidomide disaster was a factor in forming the EMA.

Thalidomide, used to prevent morning sickness in pregnant women and one of the most successful central nervous system depressive drugs at the time, was alien in Europe in 1957; by 1960, thalidomide entered more than 20 countries in Europe and Africa. It is true that the thought of approval regulatory procedures in the concerned countries was not there. Lack of regulatory affirmation to the contrary and only based on the animal toxicity and safety profile data, thalidomide entered almost all European markets. At the end of 1961, instantaneously, thalidomide was pulled off shelves because it was associated with astringent teratogenic effects and unfortunately, the accident was catastrophic, with many thalidomide-exposed babies suffering permanent disability. This is known as the thalidomide disaster. The EU learned from the disaster, and in 1965, the first European Directive was laid down known as 65/65/EEC, declaring that no medicine or product having medicinal importance could be placed on the market in an affiliate accompaniment unless approval had been issued by the competent ascendancy of affiliate member of the state. The concept was that every medicine manufacturer had to seek approval from the competent authority of the country for business in that country.

The Second European Directive (75/319/EEC) in 1975 reduced the complication complex in gluttonous approval beyond Europe by introducing mutual recognition so that approval in one affiliate country would allow business in added affiliate countries after acceptance of the absolute approval process.

The Committee for Proprietary Medicinal Products (CPMP) was formed by the directive of 75/319/EEC and the committee exchanged the views of experts of various member states of any disputed products. The action is an accomplished acquired license of a product in one particular member state and this affiliation applicated for granting the licensing authorities to another state. The other state may be consulted or clarify the specific concern to the originating country where the license was granted. Any conflict of interest or any unresolved matter is to be referred to the CPMP. The process rapidly introduced medicines into the European market without delay.

In 1987, a number of new molecules reached markets using biotechnology approaches such as recombinant DNA products or monoclonal antibodies. The biggest hurdle was the regulatory procedure of new biotechnologically derived products. The CPMP was the first appellate authority to the introduction of biotechnology-derived molecules before the normal national regulatory review. In 1992, marketing of the drug was established in a single-window system because all EU countries were under a common currency and turned into a unified economy. The birth of European Agency for the Evaluation of Medicinal Products or European Medicines Evaluation Agency (EMEA) occurred in 1995 with the incorporation of the CPMP and the Committee for Veterinary Medicinal Products. The EMEA was functionalized to review and operate the current procedures. The system is incorporated by centralized or decentralized licensing procedures. The centralized system enlisted the majority of areas of the disease and concerned medicines, and in this case, the decision of the centralized system was supreme and not required for review by individual member states. The drug classes not mandated by the centralized system were undertaken by a decentralized system. In this case, the former procedure was adopted and after approval of single-member state authority, other member states recognized the approval and granted the license to market it in their country. In 2001, for the review of the applications of developed medicines for rare diseases, the Committee for Orphan Medicinal Products (COMP) was established. The use of herbal medicinal products was also increasing in the European market, and to regulate herbal medicine in 2004, the Committee on Herbal Medicinal Products (HMPC) was established for the exchange of scientific opinions on traditional herbal medicines. The nomenclature of EMEA was changed to European Medicines Agency (with the acronym EMA) in 2004, and at the same time, the CPMP was renamed the Committee for Medicinal Products for Human Use (CHMP). The role of EMA has been crucial in the regulation of the drug industry and has imposed a number of regulations including the Sunset Clause in 2005, the drug approval is valid for three years, and that it is valid for an application to be required when a drug previously placed on the market is pulled out for further consideration of the application. The EMA accordingly considered post-marketing surveillance and other reports of particular drugs. The next year, in 2006, bills were imposed in the same conditions for the Orphan Drug Act. The EMA is considered a major regulatory authority after the USFDA and is headed by an executive director with full functioning of a secretariat of approximately 440 staff members. A strong network exists with approximately 4,000 technical and legal experts from various member states.

2.4 THE EMERGENCE OF THE JAPANESE PHARMACEUTICALS AND MEDICAL DEVICES AGENCY (PMDA)

Japanese history and culture are very rich regarding the use of medicine for major diseases. In history, the Buddhist temple of Osaka (593 CE) was considered the first national social institution and treated disease by indigenous lay healers using folk medicines of herbs for the welfare of humankind. In the 16th century, Western medicine entered Japan, and the Portuguese monk Luis de Almeida started the first European-style hospital. After European colonization of the South Pacific, Western medicine entered Japan even though traditional practices were dominant. In the 19th century, the Meiji government revolutionized the modern medicine system, or Western medicine, and made the first regulation in 1874, when the Meiji government brought a bill to limit the dispensing of drugs to drugstore owners, subsequently later known as dispensing pharmacists. In 1889, another bill limited the regulation of the prescription and dispensing of drugs. In 1943, Japan formed the first Pharmaceutical Affairs Law (PAL) which was directed to the quality, effectiveness, and safety of medical drugs and equipment, and this law was variously amended in 1948, 1960, and 1979. The PAL was regulated and enforced by the Ministry of Health, Labor, and Welfare. It was true that the thalidomide disaster forced every country to rethink adverse drug reactions and the compensation and related funds related to adverse drug reactions, and in 1987 forced the regulation of financial assistance for the adverse drug reactions and promotion of the research and development. Pharmaceutical safety and research or the organization for pharmaceutical safety and research (OPSR) or *kiko* was enforced and activated in 1994 for looking to review medical products. The Pharmaceuticals and Medical Devices Evaluation Center (PMDEC) was formed in 1997, and its function was limited to the point of modern reformation with international standard regulation of a systematic approval review system. The contribution and significance of Japan in the global medical market are negligible and the approval of any medicinal product is required to clinical trial Phases I–III by using Japanese subjects in Japan, even when a globally accepted drug was approved in other countries. In 2004, the Pharmaceuticals and Medical Devices Agency (PMDA) was formed and considered as an independent, non-governmental agency separate from the Ministry of Health, Labor, and Welfare, and the main aim was to keep pace with the international pharmaceutical market. The PMDA virtually took a major role to reform the pharmaceutical policy of Japan and the responsibilities were taken from the OPSR/Kiko. A major revolution took place in Japan in the pharmaceutical market after 2004, and further for pharmaceutical research promotion, the PMDA was transformed to become the National Institute of Biomedical Innovation (NiBio) in 2005 and the PMDA's function was limited to the more regulatory processes for review, safety, and relief operations.

2.5 CENTRAL DRUGS STANDARD CONTROL ORGANIZATION (CDSCO), AND THE OFFICE OF ITS CONTROLLER, THE DRUGS CONTROLLER GENERAL (INDIA) (DCGI)

In comparison to Japan, India received Western medicine quite a bit earlier because the British ruled in India for a long period of two centuries. The allopathic system of medicine was brought to India by the Dutch, French, Portuguese, and British. The legislature of the Indian system of the regulation of drugs and pharmaceuticals originated from the British. Before the 20th century, all modern medicines imported from foreign countries made without quality control and spurious drugs entered the market. The acute problem with drugs was the illegal trade in narcotics and the adulteration of common medicines.

In 1907, physician Dr. John Megaw observed that his malaria patients in Calcutta were failing to be cured and concluded that the patients failed to take the prescribed drug quinine. After examination, the quinine formulations were discovered to be variable in the market. Some were under-strength, and he pointed out the adulteration of the quinine when he published his results. He replicated his research in 1920 and again he assured that drug adulteration had increased proportionally in India. He also pointed out that in government, tea plantation hospitals, and dispensaries, the drugs were substandard. The issue was burning, and the matter was pointed out to both the government of India and its Medical Service.

Revision of the British Pharmacopoeia took place by the UK's Therapeutic Substances Act (1925) and committees of the British Medical Research Council, and the League of Nations—forerunner to the United Nations—prescribed the quality and use of cinchona products.

In 1928, Lt. Col. H. A. J. Gidney, a member of the Legislative Assembly, raised his voice against spurious, counterfeit, and adulterated drugs, and he suggested that India should follow the British policy and implementation of its Food and Drug Act, and Pharmacy and Poisons Act. An inquiry committee for the examination and recommendation of the regulatory process of the drug in Indian was constituted in 1930 by the government of India. The constituted committee was headed by Professor R. N. Chopra, Professor of Pharmacology, Calcutta School of Tropical Medicine, and two other members: Fr. J. F. Caius, Haffkeine Institute, Bombay, and a research chemist from Stanistreet and Co. of Calcutta. Based on recommendation, a bill was drafted and several protests were made against the bill. Later on, the bill was amended and implanted as the Drugs and Cosmetics Act 1940 and Drugs and Cosmetic Rules of 1945. The Drugs and Cosmetics Act (1940) was implemented from 1 April 1947 throughout the nation. The government amended the Drugs and Cosmetics Act a number of times to ensure the safety, efficacy, and quality monitoring of the medicine, and in 1962, the government extended the regulation to cosmetics in a provision to the same act. The salient feature of the Drugs and Cosmetics Act is the process and laid norms to regulate drugs and cosmetics regarding import, manufacture, distribution, and sale. The act is well described in the chapter, rules, schedules of drugs, etc. Also, the act directed to state government to control the manufacturing process and sales by respective control organization. Central government controlled the marketing authorization, monitoring adverse reactions to market-approved drugs and the implementation of the regulations from time to time.

The act compiles the establishment of the Central Drugs Standard Control Organization (CDSCO), and the office of its controller, the Drugs Controller General (India) (DCGI) to regulate India's drugs and cosmetics imports, manufacture, distribution, and sales. The CDSCO is under the Ministry of Health and Family Welfare, government of India, and is headed by the DCGI. The authority has designated six laboratories to control the quality of drugs, and operates zonal and subzonal offices throughout India. Further controlling export and import, the authority has seven offices throughout the nation in major ports and airports.

The Drugs Technical Advisory Board (DTAB) and Drug Consultative Committee (DCC) were constituted by this act and evaluate separately from the modern system of medicine and Indian traditional systems of medicine, respectively. Further, by the act for the quality control of the central drug-testing laboratory, Kasauli was established. The functions of the DTAB are purely technical advisory bodies and from time to time advise the central and state governments for the regulation, amendment, and technicality of the regulation. Most of the DTAB members are technical persons, whereas the DCC is implanting authority and members are from central and state drug control officials and their functioning is limited to enforcement, prosecution, and implementation of the Drugs and Cosmetics Act.

In the regulation of drugs, the Ministry of Chemicals and Fertilizers has a significant role in the implementation of industrial policy. The drugs are regulated by the direct involvement of two ministries: the Ministry of Health and Family Welfare, which is limited to public health and regulates drugs in India, and the Ministry of Chemicals and Fertilizers. Other ministries also indirectly regulate drugs in India, including the Ministry of Environment and Forests for pollution and environmental issues of the discharge of effluents and packets; Ministry of Finance for drug prices; Ministry of Commerce and Industry for the economic policies of the drugs; Ministry of Science and Technology for the intellectual property of the discovery, including the patents; and Directorate General of Foreign Trade for importing or exporting of drugs and finished products. The license and quality control authority in India is CDSCO under the Ministry of Health and Family Welfare.

Also, different laws directly or indirectly control the regulation of drugs in India, including the Poison Act, 1919; the Drug and Magic Remedies (Objectionable Advertisement) Act, 1954; The Medicinal and Toilet Preparation (Excise Duties) Act, 1956; the Narcotic Drugs and Psychotropic Substances Act, 1985; the Insecticide Act, 1968; and The Drug (Price Control) Order, 1995 (under the Essential Commodities Act).

2.5.1 The Functional Bodies of Drug Regulatory Authority in India

1. Drug Control General India (DCGI): The apex regulatory body under the Ministry of Health and Family Welfare regulates and enforces body clinical trials including overseas and in the country.
2. Indian Council of Medical Research (ICMR): This apex body ordinates and promotes biomedical research in India and provides advice to DCGI but has no direct regulatory controls for drug affairs.
3. Genetic Engineering Approval Committee (GEAC): An advisory body compromised of experts of genetic engineering and molecular biology. Its approval and recommendation are required for any clinical trials of biotech products from GEAC and directed by DCGI.
4. Department of Biotechnology (DBT): Considered the apex body for controlling the research and implementation of modern biology and biotechnology in India.
5. Atomic Energy Review Board (AERB): Regulation and approval authority for radiation equipment, including registration/commissioning inspection and decommissioning of installations.
6. Baba Atomic Research Centre (BARC): Any radiation-related projects in India are concerned with BARC and all radiation projects are controlled by BARC approval. The use of radiopharmaceuticals in a clinical trial is indirectly controlled by BARC; in most cases, DCGI refers to BARC for their expert opinion of the clinical trial for the radiopharmaceuticals.

7. Drugs Consultative Committee (DCC): The advisory body provides the expert and technical opinion to the Central Drugs Standard Control Organization.
8. Central Drugs Laboratory (CDL): The national statutory laboratory for the quality control and investigative of drugs in the market and controlled by the government of India.
9. Central License Approving Authority (CLL): The statutory and official body within the CDSCO purview is assigned responsibility for issuing "No Objection Certificates" for the manufacturing and marketing of drugs in India.

Drugs Technical Advisory Board (DTAB): The main and important advisory council composed of various technocrats and scientists to provide technical guidance to the CDSCO including the importing of drugs in India, and the function of CDSCO is limited to the advisory of Drugs Technical Advisory Board (DTAB) and the Drugs Consultative Committee (DCC), CDSCO is also responsible for the observing and medication or formulations that are sold in India are screened for adverse drug reactions (ADRs) mentoring, and reporting to WHO or USFDA as when necessary, ensuring the good manufacturing practice (GMP) certification scheme as per WHO norms, and reviewing and screening of applications and finally granting authority for "No Objection Certificates" for export of unapproved/banned drugs.

In the drug regulatory regulation, The State Drugs Control Organization is also a responsible component for providing under the central directive or along with the CDSCO for issuing for the license drug manufacturing, approval of drug-testing laboratories, approval for the drug formulations for manufacture, approval authority for a license to sell drugs, and storage of drugs. Also, the state drug control organization executed the pre- and post-licensing inspections and control the manufacturing process to watch the manufacturing process under strict GMP norms run by the respective state. The state government has power to prosecute any legal violations of the legal provisions and spurious or substandard drugs, and the Central Drugs Laboratory undertakes testing of such drugs.

BIBLIOGRAPHY

1. European Medicines Agency. Available from: www.emea.europa.eu. Accessed Mar 21, 2007.
2. Food and Drug Administration. Available from: www.fda.gov. Accessed Mar 16, 2007.
3. Jefferys DB, Jones KH. EMEA and the new pharmaceutical procedures for Europe. *Eur J Pharmacol.* 1995; 47:471–476.
4. Medical Research Council. Streptomycin treatment of pulmonary tuberculosis. *BMJ.* 1948; 2:769–782.
5. Neuhauser D, Diaz M. Shuffle the deck, flip that coin: Randomization comes to medicine. *Qual Saf Health Care.* 2004; 13:315–316.
6. Neuhauser D, Diaz M. Daniel: Using the Bible to teach quality improvement methods. *Qual Saf Health Care.* 2004; 13:153–155.
7. Pharmaceuticals and Medical Devices Agency. Available from: www.pmda.go. Accessed Mar 21, 2007.

8. Bhat A. Evolution of clinical research: A history before and beyond James Lind. *Prespect Clin Res*. 2010; 1:1:6–10.
9. Singh H. Sir Ram Nath Chopra: A profile. *J Young Pharm*. 2009; 1:192–194.
10. The Drugs and Cosmetic Act (1940) and The Drugs and Cosmetic Rules, 1945 as amended time to time. Ministry of Health and Family Welfare, Government of India.
11. Department of Chemicals and Petrochemicals (Home page), Ministry of Chemicals and Fertilizers. Available from: http://chemicals.nic.in. Accessed Dec 2014.
12. Ministry of Environment and Forest (Home page). Available from: www.moef.nic.in/index.php. Accessed Dec 2014.
13. Ministry of Finance (Home page). Available from: www.finmin.nic.in. Accessed Dec 2014.
14. Department of Industrial Policy and Promotion, Ministry of Commerce, and industry. Available from: http://dipp.nic.in/English/default.aspx. Accessed Dec 2014.
15. Department of Science and Technology, Ministry of Science and Technology. Available from: www.dst.gov.in. Accessed Dec 2014.
16. Directorate General of Foreign Trade, Ministry of Commerce and Industry. Available from: http://dgft.gov.in. Accessed Jul 7, 2012.
17. The Pharmacy Act. 1948. Available from: www.gujhealth.gov.in/images/pdf/legis/pharmacy.act_1948.pdf. Accessed Dec 2014.
18. The Drugs and Magic (Objectionable Advertising) Act. 1954. Available from: www.rfhha.org/images/pdf/Hospital_Laws/Drugs_Magic_remedies_(%20advertisement)_act.pdf. Accessed Dec 2014.
19. The Narcotic Drugs and Psychotropic Substances Act. 1985. Available from: www.dor.gov.in/sites/upload_files/revenue/files/pdf/NDPS_Act_1985_as_amended.pdf. Accessed Dec 2014.
20. The Insecticides Act. 1968. Available from: www.cibric.nic.in/Insecticides_act.htm. Accessed Dec 2014.
21. The Medicinal and Toilet preparation Act. 1985. Available from: www.dor.gov.in/medicinalandtoilet act. Accessed Dec 2014.
22. The Drug (Price Control) Order. 1995. Available from: www.nppaindia.nic.in/drugprice95/txt1.html. Accessed Jul 7, 2012.
23. The industries (Development and Regulation) Act. 1951. Available from: www.dipp.nic.in/English/policies/Industries act1951.pdf. Accessed Dec 2014.
24. The Trade and Merchandise Act. 1958. Available from: www.indiaip.com/in.dia/trademarks/acts/tmact1958.htm. Accessed Dec 2014.
25. The Indian Patent Act. 1970. Available from: www.spicyip.com/docs/statute-combined%20patents%20act%202005.pdf. Accessed Dec 2014.

3 Regulation of Herbal and Ayurvedic Drugs in the Context of AYUSH and DCGI

3.1 INTRODUCTION

India's medical facilities are expanding daily and are a popular destination for medical tourists from abroad. Additionally, at a reasonable cost, India has developed a substantial primary, secondary, and tertiary healthcare infrastructure, as well as a skilled workforce in the public, nonprofit, and private sectors.

Western drug discovery and ayurvedic herb use have different concepts. In contrast to Western thought, which holds that particular active components are responsible for the medicinal effectiveness of herbal drugs, ayurveda holds that the entire plant is valuable. According to ayurveda, the bioactivity of herbal medicines is lost when the photochemical components of the plant are separated, and the synergism of the plant constitution throughout the entire plant is what gives rise to therapeutic action. Similar to how complete meals are easier for the body to identify and digest the vitamins with separate nutritional values; the entire plant has qualities that when used in their natural form produce a stronger efficiency when all parts are operating together as a whole.

Most conventional pharmaceuticals were restricted under Drugs and Cosmetics Act of 1940 and Drugs and Cosmetics Rules, 1945. Since the Drugs and Cosmetics Act was amended in 1959, there have been concerns about the safety and efficacy of herbal medications. In 1993, a high-level expert group was established to address these concerns. The expert group proposed that, with the exception of certain uses mentioned in historical texts, no new herbal drugs other than those certified by certification regulators be permitted to be made or promoted. In India, there are two kinds of herbal pharmaceuticals that are differentiated from one another.

In the face of increasing global demand and recommendations from the WHO, the Department of Indian Systems of Medicine and Homeopathy (ISM and H) regulates the licensing of proprietary botanical medicines and new patents with regard to safety and effectiveness. As an amendment to the Drugs and Cosmetics Act of 1940 the government of India has made a notable step to adopt good manufacturing practice (GMP) into law as of June 2000 for pharmacies manufacturing Ayurvedic, Siddha, and Unani (ASU) medicines. GMP implementation raises the standard and quality of herbal pharmaceuticals while also meeting regulatory criteria for staff, infrastructure, and quality control. Registered ayurveda practitioners and educational

DOI: 10.1201/9781003398431-3

institutions were granted an exemption from GMP and were only allowed to prepare medications for their patients, not manufacture or sell them.

3.2 SAFETY CATEGORY

Any herbal drug's top priority is safety, and it is only considered safe when it poses no risk or potential harm to the consumer. The three WHO safety categories are as follow.

1. Category 1: Over a long time, safety was established.
2. Category 2: Safety aspects have been well documented and the product can be used under specific conditions.
3. Category 3: Any new substance or herbal drug is identical to the class of drugs and safety data required in this category.

3.3 CLASSIFICATION OF HERBAL MEDICINE CONCERNING REGULATORY AUTHORITIES

An ayurvedic or herbal drug is regulated by either DCGI or AYUSH, and there are certain logical distinctions in the regulatory framework. According to DCGI, allopathic drug should be used instead of conventional or herbal medicines when entering clinical trials for any treatment intervention. Depending on its source for therapeutic or clinical application, herbal medicines can be divided into four classes. Although these categories are not always mutually exclusive, they have enough distinguishing characteristics to allow for a useful assessment about the concern of safety, efficacy, and quality might be evaluated and enhanced.

3.3.1 Category 1: Indigenous Herbal Medicines

Indigenous herbal medicines are medicines that have been traditionally or historically used in a region and are well known to the local community due to long-term usage in terms of composition, therapy, and dosage. Although a local healer has used this type of traditional medicine for a very long time as folk medicine, comprehensive information may not always be available. However, the required safety and efficacy are outlined in the national regulations for herbal drugs if the remedies in this classification enter the market or travel outside of the local area or to another region of the country.

3.3.1.1 Safety Requirement

Initially, AYUSH assess the safety of the specific Category 1 herbal remedy. Safety data are not required if the medications fall under safety Category 1, but AYUSH requirements for the safety of herbal medications are necessary if the medications fall under safety Category 2, or "herbal medicines of uncertain safety"; or if they fall under safety Category 3, which is treated as equivalent to a new substance and evaluated in accordance with all applicable regulatory requirements, including a clinical trial.

3.3.2 Category 2: Herbal Medicines in Systems

The proof and comprehensive overview that medications have been used for a very long time are referenced in literature, and their therapeutic action is documented with their unique ideas and conceptions and approved by the countries.

3.3.2.1 Safety Requirement

It is necessary to review the safety of a certain category of medicine in order to determine the safety and therapeutic efficacy of herbal medicines in this category. It is not necessary to provide safety data in this instance for herbal medication safety Categories 1 or 2. The herbal medicine is classified as "herbal medicines of questionable safety" and is considered as a new chemical entity in accordance with DCGI standards.

3.3.3 Category 3: Modified Herbal Medicines

Modified herbal medicines, or Category 3, refers to Category 1 and Category 2 herbal medicines that have had their amount, kind, route of administration, therapeutic components, method of manufacture, and medical indication changed. They must meet all regulatory requirements, including those relating to safety and efficacy.

3.3.3.1 Safety Requirement

Required safety information is necessary as per novel chemical entity compounds and as natural remedies with questionable safety.

3.3.4 Category 4: Imported Products with a Herbal Medicine Base

This category includes imported herbal medicines in both their raw and finished forms. For this category of herbal drugs, the country of origin and registration are requirements. The market and elements of using herbal medicine in clinical important use should also be highlighted.

3.3.4.1 Safety Requirement

The safety and efficacy data must be provided to the national authorities of the importing nation and must satisfy the safety and efficacy standards set forth in the nation's legislation governing the use of herbal remedies.

In an Indian company or when involved in importing this category of herbal medicine, exhaustive safety data is required and must be submitted to AYUSH or DCGI. This class of medication is regarded as having "uncertain safety" and is subject to new chemical entity regulations.

3.4 DCGI'S APPROVAL

The following plant-based herbal medicines are under DCGI preview:

1. Plant material that is claimed to be employed in specific clinical or therapeutic uses whether it takes the form of an extract, isolated chemicals, or another type of plant product will be handled as a new chemical entity if this assertion is not initially made in the texts of traditional systems.

2. If the claim is that herbal drugs or herbal botanicals are just as effective as allopathic medicines, and the herbal medicine can be used as an alternative to allopathic medicines, or it can be used in allopathic hospitals, and traditional uses of the herbal medicine not mentioned, or it can be treated as a new substance or new chemical entity (NCE). The regulatory process was carried out in accordance with DCGI criteria, and it included data on acute, sub-acute, and chronic toxicity, as well as regulatory authority approval for its clinical evaluation.
3. Using this herb offers a fresh approach to medicinal use, and the user is never mentioned in classical literature. Additionally, this herb is under the NCE.
4. A therapeutic claim that has been altered and is thought to be NCE from ancient literature.
5. Any formula changes or method of deviating from classical literature is considered as NCE.

3.5 RELAXATION OF REGULATIONS REGARDING HERBAL MEDICINE

When a herbal drug is used for less than three months and the Phase I trial provides enough information to draw conclusions, both DCGI and AYUSH agreed to waive the Phase II requirements. Phase II, Phase III, and multi-centric Phase III trials for the medication are necessary because it takes longer than three months to work. Following approval by the relevant scientific and ethical committees of the involved institutes for clinical trial, all regulatory criteria are followed, and depending on the results of the Phase II trial, the Phase III trial is subsequently organized. As previously stated, all herbal drugs used in clinical trials should be produced in accordance with Schedule M of the Drug Cosmetics Act of 1940, and according to certified GMP.

A recent addition to the Act mandates that any herbal medications manufactured for any purpose related to illness detection, prevention, mitigation, or treatment be produced in accordance with the formulas outlined in the literature included in schedule I to the said act.

All herbal medications should be produced and standardized in accordance with GMP standards. According to what was previously stated, plants and herbal remedies may not be required in the Phase I study. Instead, when a herbal remedy is used for longer than three months, toxicity testing should be conducted on two species of animals for 4–6 weeks before the herbal medication enters Phase II studies. Additionally, toxicity data in two species of animals for 4–6 weeks are needed for the Phase III trial. Phase III trial and multi centric Phase III trial are approved after examining Phase II results. Clinical trials should be conducted in accordance with biomedical research's ethical standards, which include protecting vulnerable populations and disclosing risk and benefit ratios to patients with their agreement. A qualified Ayurvedic, Siddha, or Unani doctor should be the lead investigator or co-investigator.

BIBLIOGRAPHY

1. R.K. Sharma and R. Arora. Herbal drugs: Regulation across the globe. *Herbal Drugs—A Twenty First Century Perspective* (1st ed.). JAYPEE Brothers Medical Publishers (P) Ltd., New Delhi; 2006; 625–627.
2. R. Verpoorte and P.K. Mukherjee. Overview of global regulatory status. *GMP For Botanicals—Regulatory and Quality Issues on Phytomedicines* (1st ed.). Business Horizons Pharmaceutical Publishers, New Delhi; 2003; 22–27, 41.
3. R.D. Chaudhry. Regulatory requirements. *Herbal Dug Industry—A Practical Approach to Industrial Pharmacognosy* (1st ed.). Eastern Publishers, New Delhi; 1996; 537–546.
4. Guidelines for the regulation of herbal medicines in the south-east Asia region developed at the Regional Workshop on the Regulation of Herbal Medicine, Bangkok, 24–26 June 2003, World Health Organization, Regional Office for South-East Asia, New Delhi. Available at www.searo.who.int/LinkFiles/Reports _TradMed82.pdf.
5. D. Warude and B. Patwardhan. Botanicals: Quality and regulatory issues. *Journal of Scientific and Industrial Research*. 2005; 64: 83–92.
6. Drugs and Cosmetics Act. 1940. Available at cdsco.nic.in/html/Drugs and Cosmetic Act. pdf. Accessed 17 August 2009.

4 Traditional Medicine (TM) Regulatory Requirements

4.1 INTRODUCTION

Traditional medicine (TM) has gained popularity worldwide as a sustainable and cost-effective alternative to allopathic medicine. In most countries' healthcare systems, allopathic medicine plays a crucial role and TM is not recognized as an allopathic system; it is termed "complementary and alternative" (CAM). Allopathic medicine evolved from the 19th century using the proper scientific approaches. Mostly allopathic medicine is developed from herbs. Most multinational pharmaceutical companies typically spend most of their research and development budget to discover new drugs from herbal medicine. It is reported that more than US$110 billion is annually invested to explore new herbal medicine. Most pharmaceutical companies invest funds to develop drugs to fight terminal diseases, including cancer and AIDS, from herbal medicine when there is no allopathic medicine available to cure the disease. Herbs have been used since ancient times to treat all illnesses and were the mainstream in medical science from ancient times. Lack of methodology and proper scientific approach in herbal drugs becomes unconventional and plays an insignificant role. Most of the allopathic pharmaceutical industry synthesizes the active ingredients and enriches the active ingredient from the herbal source. Ancient healers prescribed the crude drug in the form of juice, powder, decoction, or infusion, and later they modified the formulation without knowing the chemical composition of these herbs. The modern techniques facilitated to isolate active ingredients rather than using the whole plant.

Rather than using a whole plant and medicinal properties of whole plants, a scientist identifies, isolates, and extracts the active properties. Plants contain other phytochemicals including minerals, vitamins, volatile oils, glycosides, alkaloids, bioflavonoids, and other substances which are equally important in supporting a particular herb's medicinal properties. Active compounds of herbals may become toxic in relatively small doses and safety aspects are most important. Previously, scientific validation of herbal drugs was a big question, but with recent advancement in instrumental techniques, clinical trials, published data herbal products gets a greater height and respect in developed and developing countries for their basic healthcare requirements due to their availability, affordability, and accessibility. According to the WHO, 80% of the world population (almost four billion people) use herbal medicine for healthcare. India is rich in biodiversity because of its different agro-climatic zones and has 20,000 plants which have medicinal value. Further, India is very rich

DOI: 10.1201/9781003398431-4

in culture including the use of traditional knowledge, and most of the plants are used by the traditional communities. With edges of DCGI and AYUSH, India classified almost 7,500 medicinal plants in Ayurvedic drugs, 450 in homeopathic drugs, 700 in Unani drugs, 650 in Siddha, and 40 in the modern system of medicine—and all are originated from herbal sources. Almost 150 pure active photochemicals have been isolated and standardized in extraction procedures which include anti-diabetic, anti-cancer, and cardiovascular drugs, etc., but a drawback which not only concerns India but perhaps the whole world is that most traditional medicine was used and developed by different communities, but no development was made to standardize and establish quality control of the herbs. The classifications of using traditional medicine were not done properly and problems were associated with using traditional medicine. Herbs are sometimes used as a food product, dietary supplements, or herbal medicines, so proper harmonization is required. A further uniform and appropriate method for evaluating traditional medicine is not established. Therefore, based on sharing experience and information, the use of traditional medicine is useful but quality control, standardization is to be improved by rigorous or controlled regulation including the certification or registration are required to traditional health practitioners (THPs). Certain issues could potentially address from THPs' insufficient experience or improper manufacturing of traditional medicines.

4.2 BEIJING DECLARATION (8TH NOVEMBER 2008)

The WHO's Congress on Traditional Medicine was attended by experts of more than 70 countries and unanimously adopted a guideline known as the "Beijing Declaration". The main aim of the congress was the role of governments for planning national policies, rules, and standards of herbal medicine and implementing thorough national health practices. Also, the responsibility was ascertained for exploring the suitable use of traditional medicine including safety and effectiveness. Also at this congress, importance was placed on focus on recognizing traditional medicine and combining such (with other things) into their national health systems. Governments will control all regulatory measures including the deciding qualification of professionals or skilled people, and approval or licensing of traditional medicine. Proper training and technology demonstration should be implemented for traditional medicine physicians, researchers, and students to strengthened traditional medicine.

4.3 CHALLENGES OF HERBAL MEDICINE

Most herbal medicines are used for self-medication, and at times, they are taken without any prescription. It is believed that herbal treatments are safe and originate from natural origins, leading many to assume that they pose no risks. Consequently, people often consume them in larger quantities than a doctor would recommend. Safety studies are essential to determine any potential interactions with conventional drugs.

Compared to traditional pharmaceuticals, herbal medications have more complicated requirements and review procedures for their safety and effectiveness. There could be a number of reasons for unfavorable effects from taking herbal remedies. Adulteration and admixing of the toxic or hazardous materials, improper labeling including claiming, and improper doses affect the quality of herbal drugs and manufacturer must be vigilant to it. Herbal drug interaction with other drugs and foods sometimes creates a problem and healthcare providers or consumers should be cautious to use herbal medicines concomitantly with other medicines. Analysis of adverse events or Adverse Drug Reactions (ADR) is critical for herbal medicines and more complicated than in the case of conventional pharmaceuticals.

Important Ayurvedic formulations are a complex combination of ingredients/constituents, and it is difficult to verify quality control of each ingredient, e.g., *Chyawanprash*. Sometimes a single herb as medicine also leads to a problem. Monitoring the quality of the raw materials used to make herbal products is a very challenging task. Implementation of GMP by the government of India under the directive of WHO and an important step for quality control of TM and included many requirements, specific steps for quality control. GMP started from the beginning of the manufacturing including the selection of herbal plants, storage conditions, and sanitation and preparation procedures. Most of TM is compound formulation whereby the number of herbal medicines is admixed, and it is virtually impossible to isolate compounds of the formulation. Single herbals contain more than 100 constituents and mixed formulation have more than a few hundred active constituents. In TM, it is believed that the therapeutic response is due to the compound formulation and the alternation of the formulation restores the therapeutic activity or produces toxicity. Isolation of active ingredients from every herb is quite impossible because huge resources of plants and time are required.

The uniform or harmonization are non-existent sometimes different countries have different uses for a same herb, such as food, functional food, nutritional supplement, or herbal medicine, and reorganization of the herb, depending on the regulations of each country in the foods and/or medicines category. Maintaining the quality of herbal medicine is a big challenge including the source of herbals and genetic variation. Other influencing factors include the agro-climatic zone or environmental conditions where the herbs are cultivated. Process variables also affect herbal medicines, including post-harvest/collection, transport, and storage. Because of this, national medical regulation agencies have a hard time defining the notion of herbal medicines, which causes confusion among patients and customers. Sometimes the herbal constituents vary with source and environmental factors and locations. No single active ingredient is responsible for the broad therapeutic effects associated with the complex formulations. Standardization of individual herbs to give a defined constituent is quite difficult to achieve in every batch manufactured.

4.4 REGULATORY AND LEGISLATIVE QUALITY CONTROL OF HERBAL PRODUCTS

Quality control plays a crucial role in the realm of herbal medicine, encompassing the oversight of both raw materials and finished products, alongside the standardization of herbal medicines. The specific goals of these regulations are to harmonize the use of a number of recognized and relevant words in TM herbal medicines around the world. Selecting and identifying the appropriate plant species for guaranteeing the effectiveness, safety, and quality of botanical remedies. Various approaches have been suggested by regulatory agencies for raw material control, as mentioned.

1. The majority of the suggestions advised taking into account the size and shape of the plant part being employed at the outset. After that, the botanicals must be evaluated both macroscopic and microscopic, and their chemical profiles must be determined. Organoleptic characterization—including the odor, taste, and surface characteristics of sensory organs—are considered important parameters.
2. Substitutes and adulterants may closely resemble the genuine material and can be detected by using microscopy and/or physico-chemical analysis adulterants. Analysis of secondary metabolites quality control is somewhat complex. Chemo-profiling using HPLC, HPTLC, and GC are used in the quality control of herbal medicine. An examination by microscopy, along with other analytical methods, is used for the identification of herbs. Spectroscopic methods have also been suggested by certain pharmacopoeias for the analysis of botanicals. Herbo-print capillary electrophoresis and DNA analysis are two recently emerging approaches for ensuring quality.

4.5 REQUIREMENT OF LICENSING OF HERBAL MEDICINE

Nowadays, both in developed and underdeveloped countries, herbal medicine plays an important role and is quickly expanding and observing a fast-growth market of herbal medications and other herbal care products. Regulators, health professionals, and the public are increasingly expressing concerns about safety, efficaciousness, and quality control. These days, medications have turned into a day-by-day need for some individuals, particularly the elderly with multiple health complications. The herbal product is manufacturing in India and registered promising growth. The market crossed more than Rs 15,000 crores in 2012 and foreign exchange crossed Rs 9,000 crores with annual growth of almost 22%. But when compared with the world market, the significance of Indian contribution is only sharing 2.5%, though India is considered to be the most eligible country where many agro-climatic zones are located. World markets are dominated by China, Germany, Switzerland, and the UK. For most terminal diseases, treatment with herbal medicine is increasing day by day. Currently, several countries including the United States and EU countries are researching exploring herbal medicine in the treatment of cancer, AIDS, and Alzheimer's/senile dementia. Moreover, the discovery process of herbal medicine is similar including India and WHO take a major role in the harmonization of herbal medicine research.

4.6 APPROACHES FOR APPROVAL OF HERBAL MEDICINE IN INDIAN CONCERN

Herbal or Ayurveda medicine are recognized alternative treatment systems by the government of India in the categories of AYUSH. Drugs and Cosmetics Act 1940 (Amendment) regulates the approval process of herbal medicines in India with the different regulatory bodies including the Department of AYUSH (Ayurveda, Yoga and Naturopathy, Unani, Siddha and Homeopathy), CCIM (Central Council of Indian Medicine) Act, research councils like ICMR and CSIR, etc. As per the Drugs and Cosmetics Act, 1940, any means of herbal drug or remedies incorporated in the modern system (allopathic) or claiming the replacement of allopathic medicine is under the purview of the DCGI and follow all regulations laid by DCGI for approval. AYUSH drugs manufactured with under the formulas described in, the authoritative books of Ayurvedic, Siddha, and Unani systems of medicine as per specified in the First Schedule of Drug Cosmetics Act amended in 1964. ASU drug application for the purpose of mitigating or preventing illness or condition either externally or internally. GMP is an important step in the manufacturing of herbal medicine and WHO harmonized almost all countries by publishing the guidelines for quality control including the safety, efficacy, and control of the adulteration of herbal medicines. The aim is making guidelines for the harmonization of all the countries and stretching of all regulatory authorities to a common platform. Moreover, the aim of WHO is to assist national regulatory authorities, scientific organizations, and manufacturers. Standardization of herbal drugs is a challenge and meeting that most of the countries are prepared the standard monograph including India, and further, WHO has also undertaken the task of preparing pharmacopoeia monographs pertaining to herbal medicines. USFDA also controls the botanical products for the issue of safety to human issues and drafted the compulsory guidelines to follow at the time of clinical trial for the herbal products.

The approval of Indian system medicine and the regulatory procedure is simple. Registration of herbal medicinal products is controlled by the state drug licensing authority of Ayurvedic, siddha, and Unani (ASU) drugs. GMP is introduced

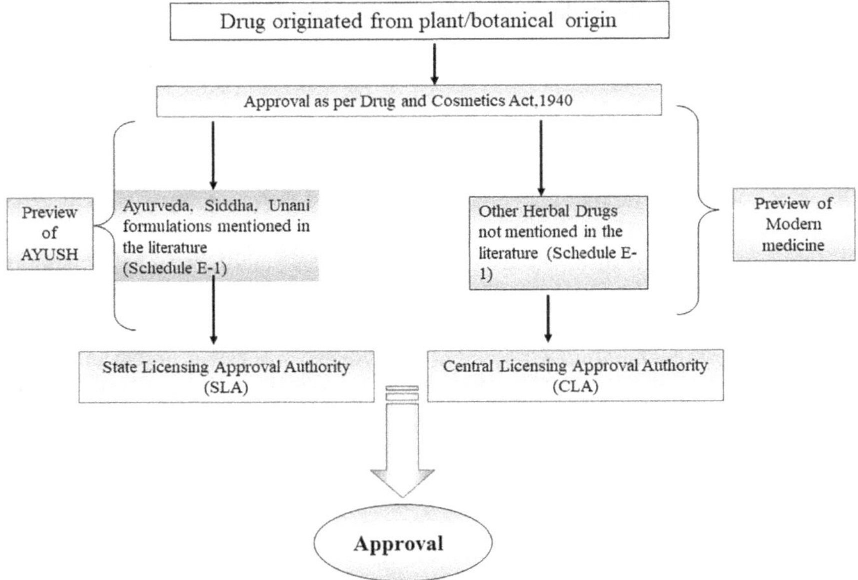

FIGURE 4.1 Herbal regulatory guidelines.

throughout India as per the Drugs and Cosmetics Acts, 1940, and the first amendment 2008, to ensure the quality and efficacy of herbal medicine. GMP controls and monitors the manufacturing process including the record of use of raw materials, machinery use, and competency by Ayurvedic or siddha or Unani licensed manufacturing units. Moreover, it is appreciated the implementation of GMP uniformly through the world, and both implementing authorities in India, where DCGI and AYUSH take the major roles. Using the excipients is controlled by Drugs and Cosmetics Acts, 1940 (second amendment 2008) or Bureau of Indian Standards Act 1986 or Prevention of Food Adulteration Act 1954 and Food Products Order for use in Ayurvedic, Sidha, and Unani drugs or the excipients mentioned in Indian Pharmacopoeia. It is true that after amending the GMP in India, the quality of the herbal product is increased, including the manufacturing process in hygienic conditions, raw material in-process quality control, proper labeling, and the process controlled by a competent person. Ministry of Health and Family Welfare (Department of Health), Government of India (MoHFW), takes the considerable step by establishing the Pharmacopoeia Laboratory for Indian Medicine (PLIM) at Ghaziabad for independent and government control testing and the reference herbarium museum.

A collaborative project between the Council of Scientific and Industrial Research (CSIR), Ministry of Science and Technology, and Department of AYUSH, MoHFW, digitized the knowledge bank of traditional medicine including the formulations, clinical use, and relevant literature in Traditional Knowledge Digital Library (TKDL) (www.tkdl.res.in). TLDL plays an important role in the literature search of traditional medicine and IPR regime, which is essential. TKDL gives protection to the Indian medicine system and provides a solid foundation in legality for the questionable inventors getting the patents by using India's traditional knowledge systems. Also, TKDL protects international patents and the international patent examiner checks the database to prevent misappropriation of Indian traditional knowledge. The database of TKDL provides more than 2.5 lakh formulations from the text of Indian traditional medicine. TKDL also provides more than 1,500 formulations from Ayurvedic, Yoga, Unani, and Sidha formulations. TKDL provides the knowledge of herbal drugs in five major international languages: English, German, French, Japanese, and Spanish.

Collaborating between the Quality Council of India (QCI) and AYUSH in 2009, certification of AYUSH product initiated for ensuring the compliance of safety and standard. Introducing GMP in the AYUSH drug is a major step toward the innovation of the Indian industry. Moreover, with the introduction of GMP, the small manufacturing industry is almost shut down. However, AYUSH relaxed some ways and certifications of standard category which ensures the compliance as per domestic regulatory requirement, whereas the premium standard certification requires a more regulatory approach. In the premium category, certification is based on compliance as per WHO and GMP requirements, and certification of levels of contaminant is also required to be mentioned. Alternatively, or most rigorously, department of AYUSH explored the possibilities of introducing a voluntary product scheme for ensuring quality and consumer satisfaction. The government of India mainly the secretary of AYUSH took the constructive step of taking a great role in the agreement with QCI in 2009 for implementing the scheme and QCI. Further, the agreement was strengthened by Multi-Stakeholder Steering Committee (MSC) under

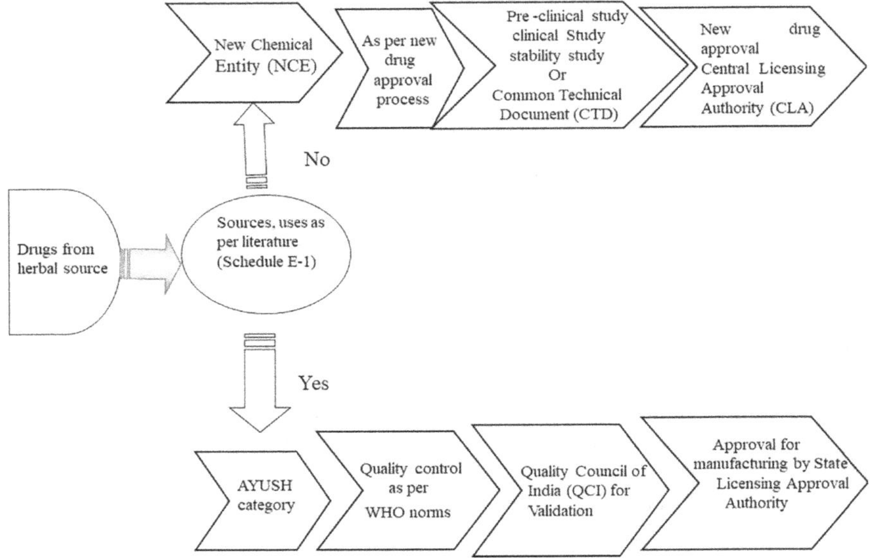

FIGURE 4.2 The process of filing of herbal drugs under purview of the DCGI and AYUSH department.

the chairmanship of the Secretary (AYUSH) with the secretariat in QCI. Various Technical Committee was also constituted by MSC and Certification Committee constituted by QCI. According to the rule with the concurrence of AYUSH and QCI, many certificate bodies (CB) were formed. Further, the CB is also accredited by the certification scheme as per ISO/IEC Guide 65, by NABCB, and/or recommended by QCI. Under the scheme, a number of private bodies were formed including Foodcert India Private Ltd., Hyderabad, Bureau Veritas Certification (India) Pvt Ltd, Mumbai, etc. WHO repeatedly pointed out that agro-climatic conditions, harvesting, and collection are responsible for the alternation of clinical values of AYUSH drugs and several guidelines introduced to regulate the quality of harvested medicinal plants. Collaboration between the National Medicinal Plants Board (NMPB) and the Department of AYUSH has prepared a guideline for the cultivation of the medicinal plants in India known as Good Agriculture Practices (GAPs) and Good Agriculture and Field Collection Practices (GACPs) guidelines. GACPs is an important guideline in the aspects of the collection of the medicinal important component from the herbs and the post-harvesting management builds between the farmer, grower, and manufacturer to maximum commercial values of herbal drugs.

4.7 CLINICAL TRIAL OF TRADITIONAL MEDICINE: A CONCEPT PROOF

In the United States, investigational new drug (IND) application of botanicals is limited only to use dietary supplements or cosmetics. Limited safety data required in an initial clinical trial of known safety of the herbal drug in dose forms and amount of dose from traditional history. Well-tolerated and known herbal drugs are typically expected to be safe particularly in Phase I clinical trials. However, there still may be some doubt or uncertainty regarding their safety. In such case, a randomized, parallel, dose-response study is considered necessary. The USFDA is more investigative in approving herbal medicine, mostly requiring the information about herbal medicine to access the safety data, such as the use of herbal medicine in other countries or the product manufacturing according to the ancient literature or based on the previous human experience. An herbal clinical trial is more complicated with the ascertained methodologies, dose selection, and additional botanical drugs. Usually, dose-response of the herbal drugs determined with placebo control with double-blind randomized clinical trials. Safety data of pre-clinical trial or Phase I or II data are required for the Phase III trial. All the clinical trial is made as per AYUSH guidelines and safety is the major issue of herbal medicine. Most issues arise at the time of standardization of the herbal medicine, extraction process, and possibilities of contamination, including heavy metals, microbiological load, and residues of mycotoxin and pesticides.

4.8 TOXIC CONTAMINANTS OF TRADITIONAL MEDICINE

Over the past decade, various reports have appeared about the negative effects of botanical medications, many of which may be traced back to the complex chemical compositions found within or on the plants themselves.

4.8.1 Microbial Contamination

Herbal products, especially those extracted using hydroalcoholic mixtures (combinations of water and alcohol), can be susceptible to microbial contamination. Microbial contamination refers to the introduction and proliferation of microorganisms such as bacteria, fungi, and molds in herbal products. The microbial load of medicinal plants is also a major concern and risk assessment and has development of Hazard Analysis Critical Control Point (HACCP) schemes has led to the increasing significance of study of microbial contamination.

Specific direction is provided for analyzing microbiological contaminations of raw and processed botanicals by several guidelines such as the WHO, British Herbal Pharmacopoeia (BHP), Indian Herbal Pharmacopoeia, and European Pharmacopoeia. In cases where the specific pesticide to which the plant material was exposed is known or can be identified by means of adequate methods and an established method for determination of that particular pesticide, the WHO recommends testing for groups of compounds rather than individual pesticides.

4.8.2 Heavy Metals

The presence of heavy metals in many herbal remedies is often covered and these heavy metals originate from contaminated soils. Maximum limits of lead and cadmium are 10 mg/kg and 0.3 mg/kg, respectively, as per the WHO. Other heavy materials including arsenic also must be mentioned.

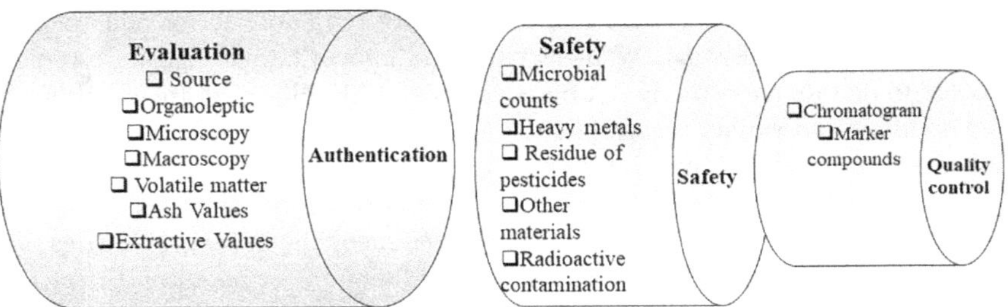

FIGURE 4.3 The standardization of herbal medicine as per WHO guidelines.

4.8.3 Radioactive Contamination

Significant risk is associated with radioactive contamination. However, radioactive contamination is gradually decreased through the manufacturing process. Therefore, no limits for radioactive contamination are proposed.

4.8.4 Pesticide Residue

The presence of pesticides in medicinal plants is a significant concern as it can pose health risks to consumers. Every country producing medicinal plant materials (naturally grown or cultivated) should have at least one control laboratory. These laboratories should be capable of detecting pesticides using specific procedures outlined in "Quality Control Methods for Medicinal Plants".

4.8.5 Plant Materials

Regardless of whether the plants are produced or naturally occurring, any country that produces medicinal plant materials must have at least one control laboratory that can identify pesticides using the guidelines outlined in quality control methods for medicinal plants.

4.9 QUALITY CONTROL OF TRADITIONAL MEDICINE

The following documents are required for quality controls of traditional medicine.

1. Description of product and clinical application.
2. Details of botanicals used in the formulation including source, organized, or unorganized drugs.
3. Previous history of use with reference and supported documents.
4. Investigational use and status to other countries.
5. Details of chemistry, quality controls, active compounds, marker compounds, manufacturing process.
6. The raw material of botanical including sources and quality controls.
7. Details of botanical drug.

TABLE 4.1
Limit of Quality Control (WHO)

Microbial load	≤ 10^5 CFU/g at the time manufacturing
Fungi, mold, or yeast load	≤ 10^3 CFU/g at the time manufacturing
Pathogenic	Free from Staphylococci, Salmonella, Pseudomonas
Coliform	≤ 10 CFU/g or free
Heavy metals	Lead and cadmium < 20 ppm
Solvent residue	Free
Aflatoxin	Free

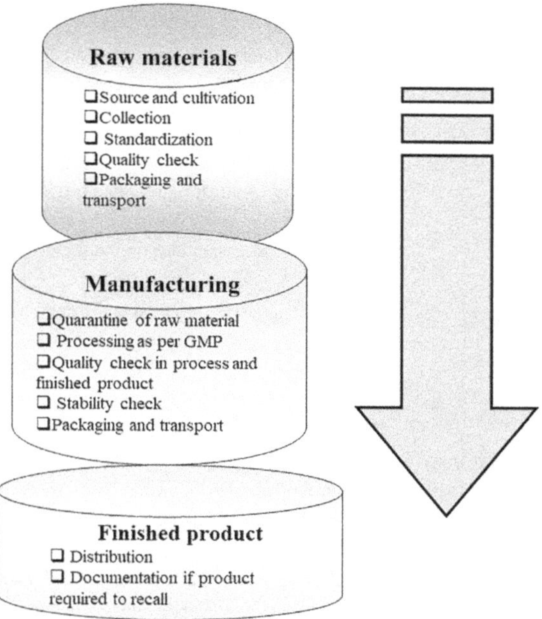

FIGURE 4.4 Good manufacturing process of herbal drugs, adopted from WHO.

8. Details of placebo control in case of the clinical trial, and the efficacy of the drugs.
9. Labeling as per Drug Cosmetics Act, 1940.
10. Environmental assessment of the herbal plants to ensuring not endangered species.
11. Details of the mechanism of action (pharmacology), drug interaction, food and drug interaction.
12. Pre-clinical toxicity information.
13. All relevant documentation for early trials for considering IND.
14. Details of clinical trial.

TABLE 4.2
Required Parameters for Herbal Drugs as per the WHO Directive

Parameters	Test
Sensory evaluation	Utilization of visual macroscopy, tactile perception, olfactory assessment, and tasting examination
Foreign matter	Includes foreign plants, foreign animals, foreign minerals, etc.
Microscopy	Consists of histological observation and measurements
Physico-Chemical Parameters	
Fingerprint	TLC/HPTLC
Ash values	Total, acid-insoluble, water-soluble
Extractive values	In hot water, cold water, and ethanol
Moisture content and volatile matter	Loss on drying (LOD), azeotropic distillation
Volatile oils	By steam distillation
Pharmacological Parameters	
Bitterness value	Bitterness in units comparable to the standard quinine hydrochloride solution
Hemolytic property	On ox blood by contrasting it with a saponin or standard reference solution
Astringent property	Tannins that bind to standard Freiberg hide powder
Swelling index	In water
Foaming index	Foam height generated under specific conditions by 1g of material
Toxicological Parameters	
Arsenic	The stain observed on the $HgBr_2$ paper is compared to a conventional stain
Pesticide residues:	Comprising of total organic phosphorous and total organic chloride
Heavy metals	Like cadmium and lead
Microbial contamination	The quantification of functional aerobic pathogens. *E. coli, Salmonella, P. aeruginosa, S. aureus. Enterobacteriaceae*
Aflatoxins	By TLC using standard aflatoxins (BI, B2, G. and G2)

BIBLIOGRAPHY

1. World Health Organization (2003). Guidelines for the Regulation of Herbal Medicines in the South-East Asia Region Developed at the Regional Workshop on the Regulation of Herbal Medicine, Bangkok, 24–26 June. Regional Office for South-East Asia, New Delhi. Available at www.Searo.who.int/LinkFiles/Reports TradMed82.pdf. Accessed 17 August 2010.
2. R.K. Sharma and R. Arora (2006). Herbal Drugs: Regulation Across the Globe. *Herbal Drugs—A Twenty-First Century Perspective* (1st ed.). JAYPEE Brothers Medical Publishers (P) Ltd., New Delhi; 625–627.
3. World Health Organization (1984). *Prepare Specific Legislation Governing the Practice of Traditional Medicine Within the Framework of National Health Legislation and Ensure Budget Appropriation to Allow the Effective Launching or Development of a*

Programme on Traditional Medicine. Report of the Standing Committee on Technical Cooperation among Developing Countries (TCDC). WHO Regional Committee for Africa, Regional Office for Africa, Brazzaville (Resolution, AFR/RC34/R8).

4. World Health Organization (2000). *Essential Drugs in the WHO African Region: Situation and Trend Analysis*. Final Report of the WHO Regional Committee for Africa, Windhoek, Namibia, 1999 (Resolution, AFR/RC49/R5).
5. World Health Organization (2001). *Promoting the Role of Traditional Medicine in Health Systems: A Strategy for the African Region*. WHO Regional Office for Africa, Temporary location, Harare, Zimbabwe (Document AFR/RC50/9 and Resolution AFR/RC50/R3).
6. World Health Organization (2004). *WHO Model Legal Framework for the practice of TM: A Bill for Traditional Health Practitioners in the WHO African Region. Tools for Institutionalizing Traditional Medicine in Health Systems of Countries in the WHO African Region*. WHO Regional Office for Africa (AFRO/TRM/2004.4).
7. World Health Organization (2004). *WHO Model Code of Ethics for Traditional health Practitioners in the WHO African Region. Tools for Institutionalizing Traditional Medicine in Health Systems of Countries in the WHO African Region*. WHO Regional Office for Africa (AFRO/TRM/2004.4).
8. West African Health Organization (2008). *Traditional Medicine Workshop Report*. Situational Analysis of the Level of Development of Traditional Medicine in the ECOWAS Member States (2007, 2008) (6). Brazzaville: The African Health Monitor WHO Regional Office for Africa.
9. World Health Organization (2005). *National Policy on Traditional Medicine and Regulations of Herbal Medicines*. Report of a WHO Global Survey. WHO, Geneva.
10. World Health Organization (2004). *Guidelines for the Registration of Traditional Medicines in the WHO African Region*. WHO Regional Office for Africa, Brazzaville (AFR/TRM/04.01).
11. World Health Organization (2009). *Handbook of Good Laboratory Practices (GLP) Quality Practices for Regulated Non-clinical Research and Development*. Special Programme for Research and Training for Tropical Diseases, Geneva.
12. World Health Organization (2004). *Guidelines for Documenting Data on Ethnomedical Evidence. Guidelines for Clinical Study of Traditional Medicines in the WHO African Region*. WHO Regional Office for Africa, Brazzaville (AFR/TRM/04.04).
13. World Health Organization (2007). *Good Guidelines on Good Manufacturing Practices (GMP) for Herbal Medicines*. WHO, Geneva.

5 Enforcement of Drugs and Cosmetics Act, 1940 to Ayurvedic, Siddha, and Unani Drugs in India

5.1 INTRODUCTION

The great India is a hotspot of biodiversity and having 12 agro-climatic zones produces a vast range of medicinal plant species to treat almost all human diseases. Ayurveda practice started almost as far back as 5000 BC and has a long history of teaching how the rationally medicinal plants can be used in healthcare. Ayurveda is a very well codified, organized, documented, affordable, and officially recognized healthcare system in India. *Vedas* (5000 BC) describes the number of plants and which can be used in the treatment of various diseases. During the era of 1000–500 BC, *Ashtang Ayurveda* (eight specialties of Ayurveda) was written, and it was believed that *Lord Brahma* conceived of Ayurveda at the same time as the universe and humanity. Also, the Siddha Medicine possesses a substantial historical lineage, while Unani Medicine traces its origins back to ancient Greece during the B.C. era. The government of India recognized whole medicine—Ayurvedic, Siddha, and Unani (ASU)—as an official treatment procedure in the Indian system of medicine. ASU medicines come under the purview of the Drugs and Cosmetics Act, 1940 and Rules, 1945. State government performs the controlling, the importation, manufacture, and sale of drugs through state licensing authorities and drugs inspectors are appointed by the state government.

5.2 CHAPTER IVA OF DRUGS AND COSMETICS ACT, 1940

Ayurvedic, Siddha, and Unani are described in this chapter of Drug Control Administration (DCA) of subsection of Act 68 of 1982, sec.2, for certain words w.e.f. 1–2–1983. The features of this act are as follow.

The Ayurvedic, Siddha, and Unani Drugs Technical Advisory Board is established under Section 33C of Act 68 of 1982. This Act is likely related to the regulation and governance of Ayurvedic, Siddha, and Unani drugs in the context of traditional systems of medicine. The primary role of the Technical Advisory Board is to advise both the Central Government and State Governments on technical issues related to AYUSH (Ayurveda, Yoga & Naturopathy, Unani, Siddha, and Homoeopathy) drugs.

DOI: 10.1201/9781003398431-5

The advisory role encompasses providing guidance on technical matters associated with AYUSH drugs. This could include issues related to manufacturing, quality control, standardization, safety, and efficacy of these traditional medicines. The functions of the Technical Advisory Board are defined to specify its responsibilities and scope of work. These functions could cover a range of activities related to technical aspects of AYUSH drugs, ensuring that the advice provided is comprehensive and specialized. The board is constituted by following persons:

1. The Director-General of Health Services (Ex-officio).
2. The Drug Controller of India (Ex-officio).
3. The principal officer u/s 30 of the act, officer related with Indian system of medicine (Ex-officio).
4. The Director Central Drugs Laboratory, Calcutta, (Ex-officio).
5. Government Analyst nominated by Central Government u/s 33F.
6. Pharmacognist to be nominated by the Central Government u/s 33F
7. One Phyto-chemist to be designated by the Central Government.
8. One teacher in Darvyaguna and Bhaishajya Kalpana (nominated by the Central Government).
9. One teacher in ILM-UL-ADVIA and TAKLIS-WA-DAWASAZI (nominated by the Central Government).
10. One teacher in *Gunapadam* as per the act u/s sec.30 (nominated by the Central Government).
11. Three practitioners from each subject of Ayurvedic, Siddha, and Unani Tibb systems of medicine (nominated by the Central Government).
12. Three representatives of each ASU drug industry (nominated by the Central Government).
13. Two members from the Ayurvedic Pharmacopoeia Committee, one member from members of Unani Pharmacopoeia Committee, and one member from the Siddha Pharmacopoeia Committee as per u/s sec 30.

5.3 CHAIRMAN APPOINTED BY CENTRAL GOVERNMENT FROM THE BOARD MEMBER

5.3.1 Functioning of the Board Members

The board members are appointed for three years with alternation of members and have following functions:

1. The board may have the delegation of the power to by-laws fixing quorums and regulating its own procedures subject to the previous approval of the central government and conduct of all business.
2. The quorum of the board should be always filled and without any vacant in the designated post.
3. The secretary of the board can be appointed, also other supporting staff can be appointed by central government.

5.4 THE AYURVEDA SIDDHA AND UNANI DRUGS CONSULTATIVE COMMITTEE (SECTION 33D)

The main function of the Consultative Committee is to advise the central government and state governments related to Ayurvedic, Siddha, and Unani medicines. Also, it is an advisory body to the technical advisory board for implementation of the act and uniform distribution throughout the country. Central government constitutes the Consultative Committee and nominates two members, whereas state government nominates one member. The meetings and decisions are taken when the central government is required to implement or amend the regulation.

5.5 MISBRANDED DRUGS (SECTION 33E)

This section defines misbranded drugs. A drug can be considered misbranded if its appearance is not aesthetic or not according to the therapeutic claims or looks as if it has excessive coating, coloration, or polish. Also, the drug is also considered misbranded if it is mislabeled or claims made of the drug can be considered as a false and misleading to the user.

5.6 ADULTERATED DRUGS (SECTION 33EE)

This section describes what is deemed to be adulterated of Ayurvedic, Siddha, or Unani drugs. Decomposition of natural product is a common and natural thing, but the decomposition should not be the negligence of the manufacturer if it is injurious to health. The drug is considered adulterated if it falls under any of the following circumstances.

1. A filthy, putrid, or decomposed drug substance.
2. Any drug contaminated with filth rendered injurious to health at the time of packaging.
3. If it contains any poisonous or deleterious substance.
4. Unprescribed or unregulated color.
5. Harmful or toxic substance.
6. Including a substance which reduces the quality or strength.

5.7 SPURIOUS DRUGS (SECTION 33EEA)

Under this section, a drug can be labeled as spurious if any of the following conditions are met.

1. The name of changes or the names are drug which are to other another class.
2. The substituted drug resembles another drug, and on the container label the name is implied to be another drug. A substituted drug resembles another drug, and the name on the container label implies it to be another drug, raises concerns related to potential issues in labeling, substitution, and potential risks to consumers.
3. If the container label states the incorrect name of the individual or company purporting to be the manufacturer of the drug.

4. Substitution of the original drug made wholly or in part by any other drug or any other substance.
5. The finished product is not considerable to a true product.
6. Regulation of manufacturer for sale of Ayurvedic, Siddha and Unani drugs (33EEB). Only standards of Ayurvedic, Siddha, or Unani can be manufactured for sale or for distribution. The allied production of the drug or substance is prescribed in relation to how that drug can be manufactured.

5.7.1 Prohibition of Manufacture and Sale of Certain ASU Drugs (Section 33EEC)

Any prohibition on the manufacture or sale of this categorized drug by the state government by notification in the Official Gazette, specify in this behalf that no person, either individually or by any other person on their behalf. It's important to note that the specifics of regulation and exemption for traditional medicine practitioners can differ widely depending on local, national, and international laws and guidelines. These regulations aim to balance the preservation of traditional medical practices with the protection of public health and safety. Therefore, the exact nature of exemptions, licensing, and registration may vary based on the jurisdiction and the specific traditions involved. Such notification can be framed as follows.

1. Manufacturing sale or distribution.
2. Adulterated or misbranded ASU drugs.
3. True list of all the ingredients displayed in the prescribed manner on the label or container. Sometimes, the patent or proprietary medicines are exempted from the disclosure of the details in the container except for the name of the active ingredient.
4. Any contravention medicines which are in the ASU drug in the provisions.
5. The importance of adhering to the regulatory framework governing ASU drugs. It suggests that any actions related to the manufacturing, storage, sale, exhibition, offering for sale, or distribution of ASU drugs must comply with the provisions of the applicable act and rules. Violating these provisions may lead to legal consequences or penalties. To get specific details, one would need to refer to the relevant act and associated rules governing ASU drugs in the applicable jurisdiction.
6. Any licenses are issued in the accordance of the ASU drug for the manufacture, for sale or for distribution.
7. Power of central government to prohibit manufacture, etc., of ASU drugs in public interest (Section 33EED). Any ASU drug which is toxic and has sustainable risk associated with the user and animals, or such drug has no therapeutic value, the power is vested in the central government to stop its manufacture, sale, and use without prejudice to the provision contained in this chapter. This includes the official procedure for giving notification to the official gazette, outlaw the production, marketing, or distribution of any medicine having acute side effects.

5.8 OFFENSES BY COMPANIES

A company has violated the act and at the time of the offence, responsible persons are counted as liable in view of the dignitaries or authority of the company. The company is also deemed to be guilty of the offense and shall be liable to be prosecuted as per laws and potentially convicted accordingly. No such provisions are rendered in this act for the person involved in committing an offense without their knowledge or the person who exercised to all due diligence to prevent the commission of such offense, but it may be that the prejudice matter in view of his involvement can be considered. The act is also silent in the subsection that the degree of offence that can result in legal action and punishment for those involved when committed offence with knowledge, complicity, or negligence.

5.9 OFFENSES BY GOVERNMENT DEPARTMENTS (SECTION 34A)

The case study of the Municipal Corporation of Delhi v. G.S. Gujaral, 1972 FAC 621 (Del) is a good example where u/s chapter IV or chapter IVA offenses are accountable to the departments of government. The central government is the charge of manufacture, sale or distribution of drugs, or where no authority is specified and deemed the responsible authority. The head of the department is responsible and deemed to be guilty under the offense and liable to be proceeded against and punished under this frame of laws, but the provisions are there if the pre-judiciary prosecution is exempted if the authority or person has the rights or sufficient proof to establish that the offense was committed without their knowledge or the person shows their due diligence without any negligence to prevent the commission of such offense.

Penalty for vexatious search or seizure is described in the section 34AA, and this section also counted as the offenses by government officers. The extended exercising powers of drug inspectors under this act or the rules made are as follows.

1. If any searches take place of a vehicle, vessel, or other conveyance without sufficient suspicious grounds.
2. Unnecessary or vexatious searches of any person without sufficient suspicious grounds.
3. If seizures of any drugs, cosmetics, articles, substances, records, documents, registers, or other material objects are done vexatiously and unnecessarily or without sufficient background of the seizures.
4. If any injury happened at the time of seizures and if the sufficient reason is not proven for the execution of duty, the injury is liable to be punished with a fine which may extend to Rs 1,000.

Publication of sentences passed under the act (Section 35) and the convicted person's or accused person's name, and place of residence and the punishment imposed upon them to be made public or in any other way the court may be published or in such other manner as the court may direct be published in a newspaper at the cost of the person.

Magistrate's power to impose enhanced penalties and the power is delegated to the magistrate as the laws of words and figures "section 32" of omitted by Act

13 of 1964, sec.29 (w.e.f. 15–9–1964). Also, the other legal proceedings including the Subs. by Act 68 of 1982, sec.38, for the Code of Criminal Procedure, 1973 is also implied. Under Act 68 of 1982, sec.38, for "any Presidency Magistrate of any Magistrate of the first class" (w.e.f. 1–2–1983), any metropolitan magistrate or any judicial magistrate of the first class are delegated the power to pass any sentence authorized by this act.

Certain offenses (36A): Trials are held under the judicial magistrate of the first class specially empowered by the state government or by a metropolitan magistrate under the sections 262 to 265 which are applicable to the trial. According to the Code of Criminal Procedure, 1973, the proved or established offenses under this act are punishable by imprisonment up to three years. Excluding the offense under clause (b) of subsection (1) of section 33I, a judge who can hand down prison terms of up to a year in length.

A legal process where a magistrate or judge reviews a trial or legal proceedings, considers the nature of the case, and makes a recorded order. If necessary, the magistrate or judge can recall witnesses who have already been examined and proceed to hear or rehear the case. This process is conducted in accordance with the applicable legal code. It is during this review and subsequent proceedings that a conviction may be framed, meaning that a judgment of guilt may be issued based on the evidence and legal arguments presented.

The specifics of this process may vary depending on the legal system and jurisdiction in which it occurs. Legal codes and procedures dictate the steps to be followed in trials and legal proceedings, and judges or magistrates play a key role in ensuring that these procedures are followed and that justice is served.

Protection of action taken in good faith (Section 37), an example is the case of Muhammad Gul v. Haji Fazley Karim, reported in the *International Law Review* (ILR) (56 Cal). Nothing in this act shall be construed to prohibit the institution of a suit, prosecution, or other legal procedure against any person for anything done or meant to be done in good faith.

The legislative process and parliamentary scrutiny of rules under a specific act, likely Act 68 of 1982. According to the specified section (Section 38) of Act 68 of 1982, rules made under the act are required to be presented before both houses of Parliament. The rules are to be laid on the floor of each house of Parliament during its session. The duration for scrutiny is mentioned as up to 30 consecutive days, and this scrutiny can span across multiple sessions.

Both houses of Parliament have the authority to vet the rules. During this process, adjustments to the rules can be made if needed. The statement suggests that if both houses concur on the rules or if adjustments are agreed upon, the rules proceed as approved. If the rules are not approved, the provision indicates that everything done under those rules before they were voted down remains legitimate. This implies that actions taken based on the rules during the period they were in force are considered valid.

BIBLIOGRAPHY

1. Subs. by Act 68 of 1982, sec. 30, for "Ayurvedic and Unani Drugs Technical Advisory Board" (w.e.f. 1–2–1983).
2. Subs. by Act 68 of 1982, sec. 30, for clause (iii) (w.e.f. 1–2–1983).
3. Subs. by Act 68 of 1982, sec. 30, for clause (viii) (w.e.f. 1–2–1983).

4. Subs. by Act 68 of 1982, sec. 30, for clauses (xi) and (xii) (w.e.f. 1–2–1983).
5. Subs. by Act 68 of 1982, sec. 31, for sections 33D and 33E (w.e.f. 1–2–1983).
6. Ins. by Act 68 of 1982, sec. 32 (w.e.f. 1–2–1983).
7. Subs. by Act 68 of 1982, sec. 2, for "Ayurvedic (including Siddha) or Unani" (w.e.f. 1–2–1983).
8. Subs. by Act 68 of 1982, sec. 33, for sections 3J (w.e.f. 1–2–1983).
9. Ins. by Act 68 of 1982, sec. 34 (w.e.f. 1–2–1983).
10. Subs. by Act 68 of 1982, sec. 34, for "of a Presidency Magistrate or of a Magistrate of the first class" (w.e.f. 1–2–1983).
11. Subs. by Act 68 of 1982, sec. 35, for "after consultation with the Board" (w.e.f. 1–2–1983).
12. Ins. by Act 68 of 1982, sec. 35 (w.e.f. 1–2–1983).
13. Subs. by Act 68 of 1982, sec. 35, for "clause (f)" (w.e.f. 1–2–1983).
14. Ins. by Act 35 of 1960, sec. 11 (w.e.f. 16–3–1961).
15. Section 33A re-numbered as section 33P by Act 13 of 1964, sec. 27 (w.e.f. 15–9–1964).
16. Ins. by Act 13 of 1964, sec. 28 (w.e.f. 15–9–1964).
17. Ins. by Act 68 of 1982, sec. 36 (w.e.f. 1–2–1983).
18. Subs. by Act 68 of 1982, sec. 37, for certain words (w.e.f. 1–2–1983).
19. The words "section 32 of" omitted by Act 13 of 1964, sec. 29 (w.e.f. 15–9–1964).
20. Subs. by Act 68 of 1982, sec. 38, for "the Code of Criminal Procedure, 1898 (5 of 1898)" (w.e.f. 1–2–1983).
21. Subs. by Act 68 of 1982, sec. 38, for "any Presidency Magistrate or any Magistrate of the first class" (w.e.f. 1–2–1983).
22. Ins. by Act 68 of 1982, sec. 39 (w.e.f. 1–2–1983).
23. Ins. by Act 13 of 1964, sec. 30 (w.e.f. 15–9–1964).
24. Subs. by Act 68 of 1982, sec. 40, for certain words (w.e.f. 1–2–1983).
25. Essential Drugs and Medicines Policy Health Technology and Pharmaceuticals Cluster, *Traditional Medicine-Growing Needs & Potential, WHO Policy Perspectives of Medicines*. Geneva: WHO; No. 2, 2002.
26. *The Ayurvedic Pharmacopoeia of India*. Chairman of the Ayurvedic Pharmacopoeia Committee, 1st ed. New Delhi: Ministry of Health & Family Welfare, Government of India, Controller of Publications; 2001.
27. WC Evans. *Trease & Evans Pharmacognosy*. 15th ed. New Delhi: Elsevier Publication; 2006. pp. 445–4.
28. https://indiankanoon.org/doc/1694683/
29. www.medindia.net/indian_health_act/drugs-and-cosmetics-act-1940-chapter-not-to-apply-to-Ayurvedic-siddha-or-unani-drugs.htm

6 Ayurvedic Drugs Regulation in India

6.1 INTRODUCTION

Good manufacturing practices (GMP) for production of Ayurvedic, Siddha, and Unani medicines are enforced by the government of India in order to ensure the quality of Ayurvedic products. The 1940 Drugs and Cosmetics Act was amended to include the new regulations, which went into effect in June 2000. The Indian Systems of Medicine (ISM) and Homeopathy Department are attempting to develop safety and efficacy standards for approving new proprietary and patent-protected herbal medicines. The draft was made in accordance with WHO recommendations, and 5,500 units are GMP compliant, with production continuing, as opposed to 2,500 non-compliant units, which have discontinued production. Ayurvedic medicine guidelines was adopted in India from 1982 and placed in Chapter IVA, Section 33-C of Drugs & Cosmetics Act, 1940. The separate drug technical advisory board for ASU pharmaceuticals is mandated by legislation to advise the government on all matters pertaining to quality assurance and drug standardization. Additionally, distinct medicine consultation groups were established to ensure that the act was applied consistently across India. These statutes gave the state the authority to license and regulate pharmaceuticals. Policy on Health, 1983, recognized that ISM medications are a helpful and effective form of treatment and incorporated the function of AYUSH in the provision of healthcare. In 2002, an independent policy was implemented for AYUSH to enhance the ISM. Additionally, national policies are being implemented to better align traditional medicine with WHO strategy.

6.2 INDIAN AYURVEDIC PHARMACOPOEIA

First Indian Ayurvedic Pharmacopoeia, which contains monographies for 80 distinct Ayurvedic medications, was released by the Indian government in 1990. Indian Ayurvedic Pharmacopoeia Part II was published in 2007 and includes new Ayurvedic medications. More than 258 different Ayurveda medications, together with the standards and quality of the plant components, are recorded in the Indian Ayurvedic Pharmacopoeia. Indian Herbal Pharmacopoeia (2002), released in 2002 by the Indian Drug Manufacturers Association (IDMA), has 52 monographs of commonly used medicinal herbs.

The release of the Indian Ayurvedic Pharmacopoeia improves the regulatory framework for Ayurvedic products. The following aim was the main concept behind the publication of the Indian Ayurvedic Pharmacopoeia.

> The Indian Systems of Medicine, specifically Ayurveda, Siddha, and Unani (ASU) drugs: The first grants excise power to the Indian Systems of Medicine as

DOI: 10.1201/9781003398431-6

an authority. This authority is empowered to publish criteria for additional plants and natural elements. Scientifically assess the efficacy of products derived from these plants and natural elements.

This indicates that the regulatory body or authority overseeing the Indian Systems of Medicine is not only responsible for setting criteria but also for ensuring a scientific evaluation of the effectiveness of traditional products.

The second emphasizes the establishment of criteria related to the excellence, authenticity, and integrity of primary substances and compound compositions used in ASU drugs. Standard Operating Procedures (SOPs) are to be formulated to ensure standardized processes in the production of ASU drugs. The creation and maintenance of a national repository of genuine reference raw materials used in ASU drug production are highlighted. This point underscores the importance of quality control, standardization, and authenticity in the manufacturing processes of ASU drugs. SOPs and a repository of genuine reference raw materials contribute to ensuring consistency and quality in the production of traditional medicines.

1. Compile monographs from the local health traditions of different communities across the nation and conduct scientific assessments of the effectiveness, safety, and quality of ASU medications.
2. Harmonization and communication with pharmacopoeia commissions and other international organizations to promote the adoption of ASU drugs globally.
3. Enforcing the ASU drug laws in accordance with Chapter IV-A of the Drugs and Cosmetics Act and Rules, which provide the quality requirements for ASU drugs.

6.3 ASU DRUG RELATES TO INDIAN LAW AND ENFORCEMENT

The following laws and regulations govern ASU drugs:

1. The Indian Medicine Central Council Act, 1970
2. The Drugs and Cosmetics Act, 1940
3. Act on Objectionable Advertisements for Drugs and Magical Remedy, 1954
4. Biodiversity Act, 2002
5. Wildlife Protection Act, 1972
6. Indian Forests Act, 1972

The Indian Medicine Central Council Act of 1970 is important for teaching Ayurveda to practitioners, and the council requires registration. The Drugs and Cosmetics Act of 1940 covers the laws and regulations governing ASU drugs. The following features are noteworthy in accordance with the rules.

1. A separate Drug Technical Advisory Board (DTAB) was established to oversee standardization and quality control of ASU medicines.
2. The Drugs Consultative Committee was established to oversee and ensure consistent application of the law in various states.
3. The licensing of ASU and drug manufacturing facilities is required, and each state's drug control agencies oversee these facilities.

4. To the benefit of the public, the central government is permitted to forbid the manufacture and distribution of specific ASU medicines.
5. Government drug analysts are divided into groups based on their education and responsibilities.
6. Enforcing ASU drug laws to designate drug inspectors and provide them authority to check drug storage and manufacturing facilities and to collect samples for quality control.
7. The statute also included provisions for penalties for manufacture, sale, etc.
8. Using a recommended and recognized approach, quality control is applied to drug testing and analysis, with an understanding of both commercial and public drug-testing facilities for sample analysis.
9. Controls the production of counterfeit, adulterated, and misbranded medications by taking appropriate measures. Schedule E medications and dangerous substances are also listed.
10. Good manufacturing practices (GMP), good laboratory practices (GLP), and labeling and packaging provisions were prioritized because of the act.

6.3.1 Guidelines for Good Clinical Practice (GCP) in Herbal Medicine

The guidelines suggest a method for conducting clinical studies of herbal medicines and medicinal plants in accordance with WHO compliance. Different herbal drugs

TABLE 6.1
Different Sechdules of Herbal Drugs as per the Drugs and Cosmetics Act

Sl	Subject	Section/Schedule
1	Technical Advisory Board for Ayurvedic, Siddha, and Unani Drugs	Section 33-C
2	The Consultative Committee on Ayurvedic, Siddha, and Unani Drugs section	Section 33D
3	Misbranded medications	Section 33E
4	Adulterated drugs	Section 33EE
5	Spurious drugs	33EEA
6	Regulation of the manufacture of Ayurvedic, Siddha, and Unani (ASU) drugs	Section-33-EEB
7	Requirements of factory premises and hygienic conditions	Schedule 1 (Rule 157)
8	Production on multiple sets of premises	Section-33-EEC
9	Ayurvedic, Siddha, and Unani medicines are prohibited from being manufactured and sold	Section-33-EEC
10	Power of central government to prohibit manufacturing, etc. of ASU drugs in the interest of the public	Section 33-EED
11	Government analysts	Section 33F
12	Inspectors	Section 33G
13	Penalties for manufacturing, selling, etc., of Ayurvedic, Siddha, or Unani drugs in violation	Section 33-I
14	Punishment for repeat offenders	Section 33J
15	Confiscation under section	Section 33K
16	Application of provisions to government departments is covered by section	Section 33L
17	Cognizance of offenses	Section 33M
18	Power of central government to make rules	Section 33N

issues as per the Drug and Cosmetics Act mentions in Table 6.1. The term "new chemical entity" (NCE) refers to an isolated photochemical that is pharmacologically active and has not previously been mentioned in classical literature. These molecules have undergone clinical trials to assess their efficacy and to ensure that they all meet all regulatory requirements.

The standards for assessing the effectiveness, safety, and quality of herbal medicines have been standardized and harmonized by the WHO. Although it is true that different rules, regulations, and laws apply in different countries, the WHO has taken the initiative to support national regulatory bodies, scientific organizations, and manufacturers in this particular sector. The WHO is crucial in the preparation of pharmacopeial monographs for herbal medicines and the guiding principles for the evaluation of herbal pharmaceuticals. Thus, to maintain the efficacy and quality of herbal medicine, regulatory mechanisms are engaged at international and/or regional controlling systems.

6.4 USFDA'S PROPOSED RECOMMENDATIONS FOR HERBAL PRODUCTS

Humans utilize herbal products extensively and unrestrainedly as traditional medicines, which may require less pre-clinical information to support early clinical findings than would be anticipated for synthetic and refined pharmaceuticals. The United States only allows the marketing of herbal medications as dietary supplements or cosmetics, and they must be registered as INDs to be supervised. Toxicological data are required before starting an early clinical trial for a product whose safety has not been established historically or in the present. Sponsors are permitted to start Phase I studies to make sure the product is safe and effective, and that the dose is considered to be suitable and well tolerated. The recommendation for a randomized, parallel dose-response research is more appropriate for determining the best dose of the product. In order to assess the safety of a novel herbal product used in a clinical trial, the USFDA needs further data (chemistry, manufacturing, and controls [CMC], toxicology, and human usage). The term "new herbal product" refers to a herbal product that has never before been marketed in the United States or anyplace else in the globe.

6.5 CLINICAL EVALUATIONS OF HERBAL PRODUCTS

Randomized, double-blind, and placebo-controlled clinical trials are typically used to test the effectiveness of herbal medications (or dose-response). Due to the generally moderately symptomatic diseases that herbal pharmaceuticals may be used to treat, as well as the fact that the active ingredient of herbal products varies from medicinal plant to medicinal plant due to agro-climatic changes, bioequivalence studies of herbal medicine are not permitted. A Phase I or Phase II study needs to have comprehensive CMC and pre-clinical safety data, whereas a Phase III clinical investigation on a botanical therapeutic product requires more specific information from the Phase I and Phase II trials. Regardless of whether the product is currently being promoted legally in the United States, abroad, or in other countries as a supplement, this additional information should be given.

All research data should adhere to accepted ethical standards of good clinical practice, including pre-informed consent and Institutional Ethical Committee (IEC) permission forms. The following information must be provided to obtain regulatory approval for an IND.

1. Product description and evidence of human use.
2. A summary of the botanicals used.
3. Use history supporting the use of plants as described in ancient literature.
4. Present investigational use.
5. Controls, manufacturing, and chemistry.
6. Raw botanical materials.
7. Drug substances made from plants.
8. Drugs made from plants.
9. Placebo.
10. Labeling.
11. Environmental evaluations or claims of categorical exclusion.
12. Information about pharmacology, toxicology, or safety.

BIBLIOGRAPHY

1. Anonymous. *The Ayurvedic Pharmacopoeia of India*. Part-I, Volume-II, Ist edition. New Delhi: Ministry of Health and Family Welfare, Govt. of India; 1999.
2. Anonymous. *The Ayurvedic Pharmacopoeia of India*. Part-I, Volume-III. New Delhi: Ministry of Health and Family Welfare, Govt. of India; 2001.
3. Arora RK, Chawla R, Shikha R, Kumar R, Sharma A, Puri SC, Sinha AK, Tripathi RM, Sharma RK. Radioprotection by Himalayan high-altitude region plants. In: Sharma RK, Arora R, editors. *Herbal Drugs: A Twenty First Century Perspective*. New Delhi: JAYPEE Brothers; 2006. pp. 301–325.
4. Chourasia OP, Parimelazhagan T, Ahmed Z. High altitude trans-Himalayan medicinal plants of defense importance. In: Sharma RK, Arora R, editors. *Herbal Drugs: A twenty First century Perspective*. New Delhi: JAYPEE Brothers; 2006. pp. 182–189.
5. Chopra RN. *Indigenous drugs of India*. Calcutta: Academic Publishers; 1992.
6. Handa SS. *Indian Herbal Pharmacopoeia*. Volumes I & II. Mumbai: IDMA; 1998–1999.
7. World Health Organization. Prepare Specific Legislation Governing the Practice of Traditional Medicine Within the Framework of National Health Legislation and Ensure Budget Appropriation to Allow the Effective Launching or Development of a Programme on Traditional Medicine. Report of the Standing Committee on Technical Cooperation among Developing Countries (TCDC). WHO Regional Committee for Africa, Regional Office for Africa, Brazzaville; 1984 (Resolution, AFR/RC34/R8).
8. World Health Organization. Regional Committee for Africa, Regional Office for Africa, Brazzaville. Essential Drugs in the WHO African Region: Situation and Trend Analysis. Final Report of the WHO Regional Committee for Africa, Windhoek, Namibia; 1999–2000 (Resolution, AFR/RC49/R5).
9. World Health Organization. Promoting the Role of Traditional Medicine in Health Systems: A Strategy for the African Region. WHO Regional Office for Africa, Temporary Location, Harare, Zimbabwe; 2001 (Document AFR/RC50/9 and Resolution AFR/RC50/R3).
10. Drugs &Cosmetic Rules. IInd amendment, Ministry of Health & Family Welfare (Department of Health), Notification; 20 January, 2005.
11. www.indianmedicine.nac.in visited in July 2014.
12. www.ccras.com visited in July 2014.

7 Licensing Procedures of the Indian System of Medicine

Ayurveda, Unani, Siddha, and Homeopathy Drugs

7.1 INTRODUCTION

The term "drug" is defined under section 3(b) of the Drugs and Cosmetics Act of 1940 as drugs for human or animal internal or exterior use, as well as meant to be used in the diagnosis, management, alleviation, or mitigation of any condition or disease in humans or animals, including preparations used to repel insects such as mosquitoes from the human body. The science-based, holistic approach of Indian medicine known as Ayurveda, yoga and naturopathy, Unani, Siddha, and homoeopathy (AYUSH) persistently addresses the problems of illness treatment.

Ayurvedic, Siddha, or Unani (ASU) drugs are produced in accordance with the formulas outlined in the authorized publications as stated in the first schedule, as per Rule 3(a) of the Drugs and Cosmetics Act (1940).

The Indian Methods of Medicine and Homoeopathy in (ISM and H) organization was founded in 1995 to promote and advance ASU healthcare systems. The Department of Ayurveda, Unani, Siddha, and Homeopathy replaced the ISM and H in 2003. In 2010, AYUSH began enforcing Rule 158 (B), which gave ASU pharmaceuticals a different dimension regarding clinical trials utilizing ASU medications and provided proof of efficacy.

The AYUSH system of medicine provides access to low-cost healthcare by drawing on traditional medical knowledge. Since AYUSH is a traditional medicinal practice, knowledge is passed down from one generation to the next. As a result, the AYUSH medical system is a tried-and-true healthcare method that differentiates somewhat from modern medicine. In contrast to the modern medical system, AYUSH medicine pioneers the regulation to comprehend the safety, efficacy, and mechanisms of action to explore globally and compete with modern medicine. The AYUSH Ministry makes a substantial contribution to the upkeep of a high standard, a logical approval procedure, and scientific confirmation for endorsing the qualities of AYUSH drugs.

DOI: 10.1201/9781003398431-7

7.2 AYURVEDIC MEDICINE

The majority of Ayurveda medications are derived from plants, and only 10% are made from metals, minerals, or merinos. Ayurveda, which has been practiced for almost 5,000 years in India, is referred to as the "Mother of All Healing". Ayurveda is based on a holistic view of the historic Vedic culture, as well as on medications.

The works of Charaka Samhita and Sushruta Samhita, written in 100 BC, contain the expertise of *Maharshi Charak* and Susrta, two well-known Ayurveda healers. These compendia—along with others like the *"Gnivesha Tantra"*, *"Bhela Samhita"*, and *"Brahma Samhita"*—are regarded as the earliest.

The *Agnivesha Tantra,* also known as the *Charak Samhita* and based on *Punarvasu Atreya's* teachings, was edited by *Charak* circa 100 BC. The *Dridhbala,* which has been revised, is regarded as one of the most recognized works of Ayurveda.

Ayurveda is built on a holistic approach; instead of using medications, it promotes the body's function by regulating energy through the fluids and nutrients that enter the cells. As a result, Ayurvedic medicines are categorized according to the movement of energy *(Vata)*, digestion of the energy *(Pitta)*, and lubrication of the energy *(Kapha)*. The dysregulation of energy—such as *Vata, Pitta, or Kapha*—is thought to be the cause of the abnormalities of body function.

The Pancha Mahabhootas are a set of environmental factors that include space *(Akash)*, air *(Vayu)*, fire *(Agni)*, water *(Jal),* earth *(Prithvi)*, and mind synergism (also known as "*Triguna*"). In addition, Ayurveda medicine overlooks the holistic approach in correlation with these environmental factors in order to maintain overall health. The use of aushadhi (drugs), ahara (food), and vihara in Ayurveda treatment is seen as the full spectrum of care, including mitigating, preventive (prophylactic), therapeutic, and curative actions.

7.3 SIDDHA

This therapeutic idea originated in India's Dravidian culture. The term "Siddha" generally refers to accomplishments and those who practice Siddha medicine are referred to as "Siddhars". In accordance with the Siddha medical system, diseases can be detected by looking at the pulse, the eyes, the tongue, the voice, the urine, the color of the body and the environment, the mental makeup, the habitat, the age, the sex, the race, the habits, the diet, the appetite, and the physical condition. The physiological makeup of the diseases is also highly dependent. Additionally, the supplementation of elements including arsenic, lead, mercury, and silver could be used in the therapy of sickness. The Siddha method is beneficial in treating a wide range of illnesses, including peptic ulcers, benign prostatic enlargement, psoriasis, rheumatic ailments, and bleeding piles.

7.4 UNANI

This medical system, which has its roots in ancient Greece, is one of the oldest in existence. This medication is governed by the AYUSH Ministry and is also referred to as "Greco-Arab" or "Unani-Tab". According to the theory of Hippocrates (460–377 BC), all sickness is brought on by an imbalance between sleep and food. This is the foundation of the medical system. This medical method was modified by numerous Greek scientists, including *Avicenna* (AD 980–1037), *Rhazes* (AD 850–932), *Galen* (AD 131–200), and others. The therapy of Unani medicine is founded on a holistic idea of prevention, rehabilitation, and cure, and the Unani system is expanded by works of *Avicenna, Al-Hawi, and Al-Qanun.*

7.5 HOMEOPATHY

The term "homeopathy" was first used in 1805 by German physician Samuel Hahnemann, derived from the Greek words *hómoios*, which means "like", and *athos*, which means "suffering".

The homeopathic system's premise is supported by experimental science as well as accurate interpretation, logical justification, and scientific construction.

7.6 NATUROPATHY

Naturopathy focuses on the five natural basic elements—earth, water, air, fire, and space—and their role in promoting health. These elements are believed to influence the balance and functioning of the body. Naturopathy is often known as "return to nature" therapy because it emphasizes harmony between the individual, community, and environment. Chronic, allergy, autoimmune, and stress-related diseases respond well to this therapy. The principles of healthy living are promoted by naturopathy, which also restores the potential of the body's innate force.

7.7 AYURVEDIC MEDICINES

The three kinds of Ayurvedic medicine are as follows:

1. Conventional medications, plant extracts, unprocessed or "raw" medications, compound formulations, and herbal mineral formulations.
2. There are several different dosage forms, including: *Avaleha/Leham/Mofak/Ilgam* (confectionary or semi-solid), *Talias/Thylam/Nei/Ghritas* (medicated oil or ghee), Shankara Sikta/Siruthugalgal (granules), *Churna/Choornam/Kvatha* Churna (fine powder), and Netrabindu/Kan.
3. Patented and privately developed medications include *Sharbat/Manappagu* (syrup), *Lepa/Malhara* (medicated wax/cream/poultice), *Kalimbu* (cream or ointment), *Avachurnam* Yoga (dusting powder), confectioneries, *Vati/Gutika/Kuligai/Marthiri/Vadagam* (tablet or pill), and suppositories.

7.7.1 Uses of Plant Parts in Ayurveda

The following plant components are employed as crude medications in Ayurvedic medicine: aerial root (A. Rt.), aril (Ar.), bulb (Bl.), dry fruit (Dr. Fr.), dry seed (Dr. Sd.), endosperm (Beeja Majja) (Enm.), exudate (Exd.), extract (Ext.), flower bud

(FI. Bd.), flower petal (FI. Petal), flower (FI.), fruit (Fr.), heart wood extract (Ht. Wd. Ext.), heart wood (Ht. Wd.), glands and hairs on fruit (G.H.F.), gall (Gl.), and inflorescence.

7.7.2 Official Monogram

The Ayurvedic Formulary Part I and Part II, which are written in Sanskrit and contain entries for the plants used in Ayurveda and their characteristics, are regarded as authoritative.

The taxonomic nomenclature is typically talked about in the introduction of the plants in the monograph, together with official names and reliable scientific literature. It is also necessary to mention the plants' further origins, popular names, scientific nomenclature and families, geographical sources, and histories. Along with the cultivation, collecting, preservation, and storage conditions, it is also necessary to discuss the macroscopical, microscopical, and sensory (organoleptic) characteristics of the substance. Purity, identity, strength, and assay method are some examples of the standardization parameters that should be mentioned. The pharmacognistic assessment and the quality control parameters, which include adulterants and substitutes, must be provided.

7.8 STANDARDIZATION OF ASU MEDICINE

To understand the quality, efficacy, adulteration, and consistency of the batch at the time of manufacture, standardization of ASU medicine is crucial.

Additionally, the uniformity of ASU medications creates a consistent remanufacturing process.

7.9 AUTHENTICATION OF ASU MEDICINE

7.9.1 Characters from Plants (Pharmacognistic)

Using a microscope, botanical (pharmacognistic) characteristics such as leaves, herbs, and flowers (whole or cut) are used to verify the authenticity of herbal medications.

Seed or fruit: The outer coat of the seed is examined under a microscope to determine its characteristics, and it must be compared to the pharmacopoeial criteria. In order to identify the stomata, such as anomocytic (irregular-celled), anisocytic (unequal-celled), diacytic (cross-celled), and paracytic, a microscopic study of the leaf drug is required. Additionally, the characteristics of the leaf—such as the stomatal index, palisade ratio, vein-islet number, and stomatal number—must be established in order to define the character and must be compared to pharmacopoeial standards.

Powder: Under a microscope, a small amount of powder and the seed coat's structure are studied, and the contents of the embryo, fixed oil, starch, and mucilage all noted. Bark is examined under a microscope to check for the presence of starch, tannins, or lignified components. The thin section of the roots and rhizomes is examined under a microscope to check for the presence of starch, inulin, lignified components, or fixed oil.

Foreign matter: Plant material should be examined using a microscope (6× or 10× magnification) or with an appropriate sieve to identify any foreign matter, such as insect or fungus waste, sand particles, or stones, and should not exceed the pharmacopoeial limit.

7.9.2 Standardization of ASU Medications

1. Determining pH values: The pH values are a negative factor in the microbial or fungal contamination-related drug deterioration. In contrast to neutral pH, fungi and other microorganisms develop more slowly in acidic environments.
2. Total ash: Total ash is a qualitative approach for quality control that can be used to determine whether there is any sand or earthy material present. Many are transformed into organic salts or inorganic salts like potassium, calcium, or oxalate when they are in the form of ash. Calcium oxalate crystals and oxide, which develop after burning, are insoluble in acid and is a qualitative indicator of the acidic matter concentration in plant roots or rhizomes.
3. Water-soluble ash: The drug's root and seed, which include tannins, sugars, mucilage, and plats acid, are particularly crucial for quality control. Water eliminates the ash that is water soluble.
4. Sulfated ash: Sulfate salts are produced by water distillation of exhausted pharmaceuticals such as tea leaves, ginger, and cloves with the addition of sulfuric acid. Regarding the air-dried medication, the proportion of sulfated ash is calculated. Guduchi's total ash and acid-insoluble ash levels, for instance, are not greater than 16% and 3%, while ginger's values are 6% and 1.7%, respectively.
5. Alcohol-soluble extractive: Nuxvomica seeds include resins, alkaloids, tannins, etc., which are used as a qualitative control of medications like asafetida.
6. Ether-soluble extractive (fixed oil content): Linseeds, mustard oil seeds, etc., contain fixed oils that are soluble in ether and are employed as a qualitative control of the medications.
7. Determining moisture composition (loss on drying): Moisture degrades chemical content proportionally and is susceptible to physical disappearance. The efficacy of crude medicines is altered by microbial deterioration and spoiling due to high moisture content. Digitalis and ergot, for instance, should have moisture contents of no more than 5% weight per weight and 8% weight per volume, respectively, and be kept in a dry environment.
8. Volatile oils, which evaporate at room temperature, give many plants their distinctive smell. This is due to their existence in drugs. Plant medicines from the Labiatae, Piperaceae, Zingiberaceae, Rutaceae, Umbelliferae, and Myrtaceae families, among others, have the property of a particular structure that secretes volatile oils, such as schizogenous or lysigenous glands, ducts, cells, or trichomes. The presence of volatile oils serves as a quality

control, for example, cinnamon (*Cinnamonium zelanicum*) bark has at least 1% volatile oil and coriander (*Coriandrum sativum*) fruit contains at least 1–4% volatile oil.

9. Organoleptic characters: The organoleptic characters are used for quality control of the plant's drugs. The organoleptic characters are determined through the visual characteristics, such as size, color, and surface characteristics, including texture and fracture of the plant drugs, while the taste and odors are also included. The aromas reveal the qualities and properties of plant-based medications. While menthol delivers the property of menthol, pungency fragrance imparts the eugenol smell of cloves. Similar to how the color offers a distinct character to any plant-based medications, cardamom, fennel, and many other plant-based drugs are all quality-controlled by color. Exhausted cloves are a lighter color.
10. Thin-layer chromatography (TLC) is a technique for qualitatively estimating the presence of any phytopharmaceutical chemicals in plant medicine combinations. Aflatoxin and trichothecene are two examples of plant-based toxins that are occasionally used to identify their presence. Refractive indices are the angle at which a light beam is transmitted and refracted as it travels from a vacuum into transparent materials or a liquid. The refractive index is a feature that all pure compounds have by nature, and it always varies when various compounds are mixed in. Most herbal medications contain oil molecules, and the fatty acids they contain are used to check the quality; for instance, castor oil's fatty acid range is 1.4758–1.527, clove oil's range is 1.527–1.535, etc.
11. Specific gravity: This is the ratio of the mass of a gas to the mass of an equal volume of air or hydrogen under specific conditions, or the mass of a solid to the mass of an equal amount of distilled water at 4°C (39°F). The oils or fats included in plant pharmaceuticals have specific gravity like cottonseed oil v.88–0.93, coconut oil 0.925, castor oil 0.95, etc. Specific gravity is another quantitative estimation parameter of specific gravity or is also known as relative density.
12. Melting point range: The temperature at which a solid transforms into a liquid. Pure herbal Phytochemicals have sharp, constant melting points, whereas the melting point range of crude pharmaceuticals is caused by the combined compounds that are presented as examples. Agar melts at 85°C, beeswax melts at 62–65°C, and wool fat melts at 34–44°C.
13. Determining the congealing range: This is the temperature at which phytochemicals begin to crystallize and is useful for qualitative analysis of the drug purity of plants. The impure products can be found in a variety of temperatures, despite the impure compound having a well-defined freezing point. When the oil content is 0.5% or less, the melting range for sunflower seed wax is 74–80°C.
14. Boiling point: The temperature at which a liquid turns into a vapor; the liquid's vapor pressure is equal to the liquid's surrounding pressure. Pure phytopharmaceuticals have a sharp boiling point, while natural components have a range of boiling points. Refined coconut oil has a boiling point of

232°C, while unrefined coconut oil has a boiling point of 177°C. Castor oil has a boiling temperature of 313°C.

15. Optical rotation: A plant-based product with an enantiomer of a chiral chemical that may change the direction of plane-polarized light can be investigated for quality control purposes. Dextrorotatory (d) or (+) enantiomers of such plant products rotate slightly to the right or clockwise, whereas levorotatory (l) or (−) enantiomers of the other plant products rotate somewhat to the left or counterclockwise. Examples include lemongrass oil from −3 to +1°, and eucalyptus oil from 0 to −10°. A mixture of sugars with each sugar having a unique optical rotation—for example, fructose has an optical rotation of −92.4°, glucose has an optical rotation of +52.7°, and sucrose has an optical rotation of +66.5°, but maltose has an optical rotation of 130.4°—is another traditional example of how to check the quality of honey. Honey's optical rotation is typically between +3° and −10°.
16. Viscosity: Measuring viscosity is crucial for plant-based medications that contain carbohydrates such as mucilage, gum, and others. When these carbohydrates are swollen by water, they create resistance to flow, which is referred to as viscosity. For example, Gond Katira (Tragacanth) at 1% concentration has a viscosity of 250–340 centipoises, while Babool (Acacia arabica) at 1% concentration has a viscosity of 3000–3800 centipoises.
17. Solubility factor: The plant's compounds have certain solubility in certain solvents, and when an unrelated solvent is combined with a soluble mixture, turbidity results. For instance, castor oil mixes nicely with half the volume of light petroleum and produces turbidity when combined with water. Another illustration is the solubility of colophony in light petroleum as opposed to that of Peruvian balsam in an equivalent volume of alcohol. In alcohol, adulterated plant components are more soluble. The majority of alkaloids are insoluble in water or polar solvents, but their salts are soluble in both.
18. Value of iodine: Oils extracted from plants primarily comprise triglycerides, which are long chains of fatty acids with striation or unsaturation at the carbon–carbon double bonds. One distinctive quality of a plant's products is how saturated or unsaturated they are, and if any of the products are contaminated, the saturation levels will change. The iodine values of groundnut oil, olive oil, palm oil, and palm kernel oil, if unsaturation is present, are 86.00 g I_2/100g, 81.00 g I_2/100g, 53.91 g I_2/100g, and 36.74 g I_2/100g, respectively.
19. Acid value: Plants include acidic hydroxyl radicals, which are primarily acidic and utilized to inspect the vegetable oil for quality. The amount of potassium hydroxide (KOH, in milligrams) needed to neutralize 1 g of oil containing free fatty acids is how the acid value is stated. Olive oil has an acid value between 6.6% and 7.4%, while coconut oil has a value between 22% and 27%.
20. Peroxide value: Peroxide value is a warning sign of impending oil or fat breakdown. In this test, iodide is utilized, and the fat or oil peroxide is reacted with to release the iodine ions, which are then measured using sodium thiosulphate. Fats with peroxide values under 10 are regarded as

high-quality oils, while those with peroxide values above 10 are deemed rancid, as in coconut oil 15 is the maximum allowable peroxide value.

21. Non-saponifiable matter: Unsaponifiable matter is comprised of non-hydrolysable elements, and practically all vegetable oils have less than 2% of it. It is a sign of purification or the addition of any impurity, fat, or oil.
22. Mineral oil detection (Holde's test): Codfish liver oils' unsaponifiable values range from 180 to 192, while refined coconut oils' unsaponifiable matter exceeds 15. There should be less mineral oil and more vegetable oil, oil from crude medicines, or edible oil. When oil is combined with the hydroalcoholic solution and oil forms into soap, mineral oil can be recognized by the turbidity that results.
23. Rancidity test ***(Kreis*** Test): Rancidity is the oxidation of free fatty acids, which primarily affects numerous natural substances like ghee, butter, butter oil, etc., and compromises their nutritional value. Phloroglucinol in diethyl ether solution, which becomes pink when rancid oils are present, is used to determine the degree of rancidity. At 234 nm and 268 nm, the *Kries* test reveals features of rancidity.
24. Extractable value (alcohol and water): Extractable matter is a quantity measure for evaluating the components of crude drugs. Water-soluble extractive values are used for water-soluble components like tannins, glycosides, mucilage, etc., while alcohol-soluble extractive values are useful for glycosides, tannins, resins, etc., and ether-soluble extractive values are useful for drugs containing volatile components and fats.
25. Heavy metals (lead, cadmium, mercury, and arsenic): Because of the cultivation and environment, the majority of crude pharmaceuticals are polluted with heavy metals. The maximum allowed levels of heavy elements are those listed in the ASU Pharmacopoeias, which are lead 0 ppm, mercury 1 ppm, cadmium 3 ppm, and arsenic 3 ppm, respectively.
26. Microbial load: According to the ASU pharmacopoeias, bacteria like *Staphylococcus aureus*, *Salmonella sp.*, *Pseudomonas aeruginosa*, and *Escherichia coli* (*E. coli*) are not allowed to be present in crude medications. Additionally, the topical application was limited to 107/g and the total microbial plate count (TPC) to 105/g, while the total yeast and mold were each limited to 103/g.
27. Aflatoxins (Bi, B2, G1, G2): The ASU pharmacopoeias aflatoxins limits as prescribed in aflatoxins B, –<2 ppb; aflatoxins B, +B2 +G, +G2 <5 ppb may be observed. Herbal components, DDT (dichlorodiphenyltrichloroethane), carbamates, organophosphates, polychlorinated biphenyls, and other chlorinated hydrocarbons may contaminate crops, especially those that are grown for agricultural purposes.
28. Pesticide residue (Organo-Chlorine and Organo Phosphorous): For permissible amounts of pesticide contamination of herbal constituents, limit tests are required.
29. Total ash content: Total ash refers to the amount of after-burning residue that was produced. Sand and silica are counted as insoluble matter after the residue is diluted with 0.1 N hydrochloride.

7.10 STABILITY STUDIES OF ASU DRUGS RULE OF 161-B

According to the G.S.R. 789(E) amend the Drugs and Cosmetics Rules, 1945, was published in sub-section (1) of section 33N of the Drugs and Cosmetics Act of 1940 (23 of 1940) in Part II, Section 3, Sub-section (i) of the Gazette of India, Extraordinary, vide number G.S.R 897(E), dated the 24 November 2015 where this rules known as Drugs and Cosmetics (5th Amendment) Rules, 2016. As amended, the rule 161-B confers the stability or shelf life studies of the herbal drugs specified by the notification GSR 764(E) dated 15 October 2009, and these amended rules are applicable for all patented and proprietary Ayurvedic medicines, both new and existing products.

According to the 161-B, the expiry date of all ASU medicines shall be noticeably displayed on the label of container or package, and no medicine shall be stored or marketing or distributed the rules strictly applicable to all the new license or renewal license from the date of implemented of the notification; i.e., after 24 November 2015 while the *post facto* or earlier approved before notification of any Ayurvedic, Siddha and Unani medicines are not applicable. The license is required from State Licensing Authority for the manufacturing of ASU medicines defined under clause (h) of section 3 of the Act and shelf life or date of stability is required to submit.

As earlier said that according to rule 161-B, the shelf life to be mentioned in the label and without the mentioning of the specific storage conditions, it is assumed that the product shall be stored at room temperature which is equivalent to below 30°C.

Stability studies provide the effects of light, humidity, temperature, and the influences of the quality of ASU medicines storage conditions. The determination of stability or shelf life assessment could be done by two methods. The first method is the "cross-sectional approach" whereby the batch is selected during the last five years of the manufacturing process and the sample is withdrawn every six months for simultaneous estimation. Second, the batch selected the stability experiments are conducted under accelerated settings, while concurrently conducting stability studies with actual samples. The expiry date or shelf life may be determined in this approach based on the evaluated data.

7.10.1 Storage Condition

The International Council for Harmonization of Technical Requirements of Pharmaceuticals for Human Use (ICH) provides guidelines for the pharmaceutical industry to ensure the quality, safety, and efficacy of medicinal products. One of the key guidelines is the ICH guideline on stability testing, which includes considerations for different climatic zones.

As you mentioned, the world is classified into four climatic zones, from I to IV, in the context of stability testing. These climatic zones are defined based on temperature and humidity conditions, which can significantly impact the stability of pharmaceutical products. The guidelines provide recommendations for testing under conditions that simulate the effects of these climatic zones.

Climatic Zone I: This zone represents temperate climates, characterized by mild temperatures and moderate humidity.

Climatic Zone II: Zone II includes Mediterranean and subtropical climates, with higher temperatures and moderate humidity.

Climatic Zone III: Zone III covers hot, dry climates, where temperatures can be high, and humidity is relatively low.
Climatic Zone IV: This zone represents hot and humid tropical climates, with both high temperatures and high humidity.

The stability testing conditions specified in ICH guidelines, particularly in Volume II, Chapter 16, involve subjecting pharmaceutical products to conditions that simulate the climatic zone where the product is intended to be marketed. This testing helps assess the impact of environmental factors on the stability of the product over time.

Stability testing is crucial for establishing the shelf life of pharmaceutical products, ensuring that they remain safe and effective throughout their intended use. It involves monitoring various parameters, such as the degradation of active ingredients and changes in physical characteristics, under accelerated and real-time conditions. Pharmaceutical companies follow these guidelines to comply with regulatory requirements and to ensure that their products meet quality standards across different regions with varying climatic conditions. The shelf life (expiry date) of a product is determined by several factors, including its type, contents, and packaging material etc. Stability studies are done in two conditions. Accelerated stability studies are done at the storage temperature range 40 ± 27°C and maintain the relative humidity 75 ± 5% for six months while long-term studies are done at temperature range 30 ± 2°C and maintain relative humidity 60 ± 5% for 12 months.

The other stability studies are allowed as per the suitability of the product and proper justification. Temperature-sensitive herbal drugs are stored in lower temperatures.

7.10.2 Selection of Batches

A minimum of three primary batches are engaged for stability testing which shall be similar to the same formulation intended for marketing.

All three primary batches shall be manufactured in a pilot batch which is 10% of the commercial batch size; every year, a minimum of two batches are examined and accordingly, if any products monitor for five years for stability studies, this means ten samples shall be withdrawn. The material quality is to be the same as the batch process and different batches are used for the stability studies. The stability studies should be carried with the container along with the packaging for evaluation of storage and distribution except for bracketing and matrixing of stability studies.

7.10.3 Specification

The endpoint or marker of the stability studies is determined by the storage conditions which may influence the quality and safety of the products and accordingly the end parameters are selected. Quality means the attribution toward the physical, chemical, biological, or microbiological attributions.

The specification indicates is a list of tests, reference to analytical procedures, and proposed acceptance criteria, while the validated stability study indicates the accuracy of the test to conclude the results. The endpoint of the stability testing is according to the probable changes of the products during the long storage which includes the physical parameters like appearance, shape, odor, color, and the chemical parameters like TLC values, color reaction, weight variation, bulk density, disintegration, extractive values, and estimation of active or marker or category compounds.

The limit of the acceptance of the stability studies according to the official monographs or pharmacopoeias as specified, or else any accepted released specification whereby the test and validation are well described.

It is important to conduct stability studies to determine accepted release specifications for ASU drugs. The endpoint of shelf life is justified based on these studies, particularly when changes in storage conditions can affect the quality and safety of the products. This process is essential for regulatory compliance and ensuring the safety and efficacy of ASU drugs throughout their shelf life.

Any deviation of the acceptance criteria from the stability studies to be justified and any plant drugs using any anti-microbial preservative for enhancement of shelf life should be demonstrated during the stability studies intended for marketing.

7.10.4 Frequency of Testing

This testing schedule is in line with good manufacturing practices and regulatory requirements to verify that AYUSH drugs meet their shelf life claims and are safe and effective for consumers throughout the specified period. Testing is essential to confirm that the product remains within acceptable quality standards, and any deviations from these standards can prompt necessary adjustments or withdrawals from the market to ensure consumer safety. The ASU drugs that claim to have a long shelf life long time shall be tested every six months for one year, and afterwards annually.

Accelerated storage stability assays are a common practice in the pharmaceutical and related industries to gain insights into how products will perform under various storage conditions over their expected shelf life. The results of these assays can inform decisions about shelf life claims, product quality, and appropriate storage recommendations.

7.10.5 Evaluation Criteria

Testing requirements for Ayurvedic, Siddha, and Unani (ASU) drugs, specifically focusing on the evaluation of three batches and criteria for significant change during testing. All ASU drugs are required to undergo testing, and a minimum of three batches is typically tested. This is a common practice to ensure consistency and quality across batches. The testing process includes evaluating the complete batches of ASU drugs, including the proposed levels of active compounds, and the packaging materials. This comprehensive testing helps ensure the safety, efficacy, and quality of the products. The testing process involves monitoring for significant changes in ASU drugs. This assessment is conducted under real-time or accelerated storage conditions to simulate different environmental factors that the drugs may encounter. "Significant change" is a categorical term defined in this context. The criteria for significant change are specified, particularly in terms of the variation of claimable markers and the active compound's assay value. For claimable markers, a ±20% change from the initial assay value is considered within an acceptable range. For the active compound, a variation of ±15% change from the initial assay value is deemed acceptable. The changing of the physico-chemical parameters like moisture, ash, particle size, etc., Examples of Ayurvedic Medicine and shelf life claims mentioned in Table 7.1. within the limit of 25% of the initial value and the form of a new spot or complete disappearance of an existing spot TLC analysis are observed when stored at less than 10°C. Further, changing any physical attributes—like phase separation, color, hardness, caking, or failure to recognition of the parameters as described in the individual monographs or specification—are also addressed in the clause of significant change.

7.11 PRE-CLINICAL TOXICITY

Pre-clinical evaluation is also considered for the ASU drugs to establish the drug safety, and a study was conducted to evaluate the potential adverse pharmacodynamic effects of medications on physiological processes in relation to conditions within the therapeutic range.

The non-clinical safety evaluations are based on whether any toxic effects are evident with any organ with respect of dose dependence, relationship to exposure of the ASU drugs. The guideline issued for the safety studies as per the schedule Y, GCP guideline or the Organization for Economic Cooperation and Development (OECD)

TABLE 7.1
A Few Examples of Ayurvedic Medicine and Shelf Life Claims

Sl no	Formulation Type	Shelf Life Claim (years)
1	*Anjana* made from *Kasthaushadhi*	1
2	*Anjana* made only from *Rasa/Uprasa/Bhasma*	3
3	*Arka*	1
4	*Asava Arista*	10
5	*Chuma, Kwatha Chuma, Lepa Chuma, Danta Manjan* (Chuma)	2
6	*Dravaka, Lavana, Kshara*	5
7	*Guggulu*	5
8	*Gutika* or *Vati* containing *Kasthaushadhi* along with *Rasa/Uprasa/Bhasma/Guggulu* (including *Lepa Gutika* and *Ghan Vati*)	5
9	*Gutika/Vati* containing only *Ras/Uprasa/Bhasma* except *Naga, Vanga* and *Tamra Bhasma*	10
10	*Mandura-Lauha*	10
11	*Kupipakva Rasayana*	10
12	*Malahar*	3
13	*Naga Bhasma, Vanga Bhasma* and *Tamra Bhasma*	5
14	*Sattva* (derived from medicinal plants)	2
15	*Sharkar/Panak/Sharbat*	3
16	*Rasayoga Containing Rasa/Uprasa/Bhasma* along with *Kasthaushadhi/Guggulu*	5
17	*Shveta parpati*	2

TABLE 7.2
A Few Examples of Siddha Medicines and Shelf Life Claims

Sl no	Formulation Type	Shelf Life Claim (Years)
1	*Curanam: Kutinir CuranamAdai Curanam/Kanchi Curanam/Utkali Curanam/Pittu Curanam/Podithimirthal Curanam/Podi/Pattru Curanam/Pottanam* or *Kizhi Curanam/Ottratam Curanam/Vethu Curanam/Pugai Curanam/Kali Curanam/Thuvalai Curanam*	2
2	*Rasa-Paadana Marunthugal* (All mercurial preparation): Containing *Mooligai* ingredients along with *Thathu Porutkal/Parpam/Centuram/Cunnam/Kattu/Kalanku*	2
3	*Parpam/Centuram*: containing *Mooligai* ingredients with *Thathu Porutkal/Parpam/Centuram/Cunnam/Kattu/Kalanku* (eg. *Aya Centuram*)	10
4	*Kulampu: Araippu Kulampu* (e.g., *Agathiyar Kulampu*)	5
5	*Meluku: Idippu Meluku* (e.g., *Rasa Gandhi Meluku/Idi Vallthi Meluku*) *Mooligai Meluku* (e.g., *Malaikudara Meluku*)	3
6	*Karpam: Mooligai Karpam* (e.g. *Karisalai Karpam, Thiripala Karpam*)	2
7	*Satthu: Satthu* derived from *Mooligai* (e.g., *Seenthil Satthu*)	2
8	*Satthu* derived from Jeeva Porutkal (e.g., *Sembu Satthu* derived from *Poonagam, Mayiliragu*)	5
9	*Ilakam/Legiyam/Iracayanaam*	3
10	*Ney/Ghirutham/Kadugu*	2
11	*Panda Vaippu*	10
12	*Tailam/Ennai/Poochu* three years	3
13	*Tiravakam* (derived from *Thathu Porutkal*)	2

TABLE 7.3
A Few Examples of Unani Medicines and Shelf Life Claims

Sl no	Formulation Type	Shelf Life Claim (Years)
1	*Arq (except Arq-e-Ajeeb)*	1
2	*Arq-e-Ajeeb*	5
3	*Jawarish*	4
4	*Majoon/Dawa*	3
5	*Raughaniyat/Tila*	3
6	*Sharbat/Sikajabeen*	2
7	*Tiryaq*	3
8	*Sufoof (without salt)*	2
9	*Tiryaq*	3
10	*Ney/Ghirutham/Kadugu*	2
11	*Jawarish*	4

guidelines. The pre-clinical toxicity data are used for the dose-ranging study and initial dose section to the human clinical trial. Please refer the pre-clinical safety studies described in Chapter 139.

7.12 CLINICAL TRIAL

The clinical trial of the ASU drugs has some differences from the traditional clinical trial, though the aim is to establish the safety and quality of the ASU drugs. The clinical trial of the ASU drugs addresses the right design, the complexities, and the diverse concepts of the clinical trial. The fundamental of the ASU drugs clinical trial is more toward the holistic systems to Ayurveda approaches and acknowledged the complex system of the drugs discovery including those related to safety, dosage forms, delivery systems, clinical efficacy, and drug interactions. The epistemology of Ayurveda is a rich and the from ancient times; ASU drugs follow the holistic way to established the efficacy with tangible evidence of the medicine and validated with the vital fundamentals of the Ayurveda. The validation of therapies and techniques can be achieved by the adoption of a holistic strategy that integrates the principles of Ayurveda and bio-medicine, while ensuring the preservation of the essential principles of both systems. The *Darshanas* (doctrines of philosophy) form the basis of research and the concept of the Ayurvedic texts, i.e., *Samhitas* is generated for the Ayurveda testing like *Aptopadesa* (evidence based on therapeutic leads), *Pratyaksha* (direct evidence), *Anumana* (logical inference), *Upamana* (analogy: comparative/ controlled design) and *Yukti* (reproducible experimental evidence). *Samhitas* are the directive of the clinical trial of Ayurveda drugs and mainly two Samhitas are the fundamental of the Ayurveda of clinical trials.

First, *Sushruta Samhita* describes how the approaches to the clinical trial (which includes the observation, intervention, testing, and system validation of the trial) and the details are like *Dristaphalatwat* related to the observational design, *Tatlingatwat* related to the feasibility of test intervention, *Kriyakala* related to testing an intervention, and *Agamaatsya* related to interpretation of observations in the context system validation and reverse pharmacology or reverse renovation approach of contemporary period.

Second, *Maharshi Charak* approached the way to understand the Ayurveda drugs through *Budhipashyati-ya-bhavaan* or study through multi-dimensional approaches in planning and feasible ideology/hypothesis assessment and *Yukti-Yojna* or appropriate design through the *Yuktihtrikalah* or perpetually valid outcome and reproducibility.

7.13 PHASES OF CLINICAL TRIAL FOR AYURVEDIC DRUG/PATENT OR PROPRIETARY MEDICINES

7.13.1 Standardization and Quality Control of the Trial Drugs

The drug should be as per the pharmacopoeial standard or in-house standard to be used in clinical trial. Further, the drugs shall be compliant with the other physical conditions like the details of the raw drugs and the certification of analysis for both raw materials and final products is essential, along with the SOPs for preparation, which should be properly documented.

7.13.2 Safety/Toxicity Studies

Important aspect of Ayurvedic drugs—the need for safety and toxicity studies. Indeed, assessing the safety profile and potential side effects of Ayurvedic products is crucial, especially considering that these products often contain natural ingredients.

Ayurvedic drugs typically include a variety of natural ingredients such as herbs, minerals, and animal products. While these substances are considered traditional and have been used for centuries, their safety must be rigorously evaluated, as the presence of natural ingredients does not guarantee safety. Safety studies aim to assess the potential risks associated with the use of Ayurvedic drugs. This includes evaluating the safety of individual components as well as potential interactions between ingredients.

Toxicity studies are conducted to determine the levels at which substances become harmful. It's essential to understand the potential toxic effects of Ayurvedic drugs, especially when taken in recommended doses or over an extended period. Regulatory bodies, such as the Ministry of AYUSH in India, often establish guidelines for safety and toxicity studies for Ayurvedic drugs. Manufacturers are required to comply with these regulations to ensure the safety of their products.

Quality control measures are crucial in ensuring that Ayurvedic products are manufactured consistently and meet established safety standards. This includes testing the purity and potency of ingredients. Conducting well-designed clinical trials is a critical step in assessing the safety and efficacy of Ayurvedic drugs. Clinical trials provide scientific evidence regarding the product's safety in human populations.
Monitoring Adverse Events:

Ongoing monitoring of adverse events is important post-marketing. This helps identify any unexpected side effects that may not have been evident during pre-marketing studies.

Public awareness campaigns are often initiated to educate consumers about the proper use of Ayurvedic drugs and potential risks. This encourages responsible use and reporting of adverse events.

Ensuring the safety of Ayurvedic drugs is essential for public health and confidence in traditional medicine. Researchers, manufacturers, and regulatory authorities play pivotal roles in conducting studies, setting standards, and monitoring the safety of Ayurvedic products throughout their lifecycle. These phases are critical to demonstrate the product's quality and efficacy, adhere to regulatory requirements, and ensure that the product is safe for patients.

The specific design and objectives of clinical trials for Ayurvedic drugs or similar traditional medicines may be influenced by the nature of the product and the regulatory framework in the region where the trials are conducted. It's important to consult with regulatory authorities and follow established guidelines for conducting such trials.

7.13.3 Therapy/Procedure

Panchakarma/other therapy trials need to consider many aspects while designing the study. The deployed persons must be suitable and trained for performing the therapies, and procedures, and familiar with the diseases. Pre- and post-therapeutic procedures, duration of the treatment, disease severity, and possible errors may require monitoring. Further, the requirement must be met to consider the lifestyle including dietary habits before, during, and after performing therapy and the quality control of the medicine.

7.13.4 Phase I: Human Pharmacology

The safety and tolerability of Ayurvedic or patent or proprietary drugs is evaluated in limited volunteers. In this phase, the aim is to establish how the safety parameters of the drugs rather than establishing the safety of the drug. Schedule E-I (mentioned in what follows) of the Ayurvedic drug or patent or proprietary medicines with probable toxicity are to be studied in participants. The two main observations may be concluded.

7.13.4.1 Maximum Tolerated Dose

The tolerated dose or the dose range periods are to be studied and the nature of adverse reaction in clinical studies further any adverse reaction are to be studied in the selected dosage range. The main aim of this type of study is to determine whether toxicity is dose dependent and the range of dose whereby no toxicity or adverse reaction is found.

7.13.4.2 Early Measurement of Drug Activity

This approach reflects the typical progression in clinical drug development, where safety is the initial focus in early stages, and as the drug advances through phases, efficacy and therapeutic benefits become increasingly attention to the investigation. The early clues of beneficial effects can guide the decision to continue with the drug's development in later phases.

7.13.5 Phase II: Therapeutic Exploratory Trials

The establishment of the effectiveness of a particular indication in participants of an Ayurvedic drug/patent or proprietary medicine for a particular indication is the common concern and the side effects and risks could shortly be evaluated.

This stage is the preparative stage for the approval process of intending the marketing for the Ayurvedic drug/patent or proprietary medicine. The studies in Phase II are conducted to the homogeneous population, and such study is closely monitored.

This phase is corrective of the dose selection and is used for the Phase III trial with a large number of volunteers and provides the adequate study for the marketing of the Ayurvedic drugs to the market. In this phase, the dose selected is usually less than Phase I maximum tolerated dose. Further, this phase is used for the end point study of the concomitant medications and target populations for further studies of Phase III. Thus, Phase II study is an exploratory trial to understand the basis of dose selection and further preparedness for Phase III trial.

7.13.6 Phase III: Therapeutic Confirmatory Trials

Phase III studies are used for the confirmation of therapeutic benefits of the molecules and whether they are safe for consumption. Usually, Ayurvedic medicine is a long process and Phase III trials establish how beneficial consumption of the drug is. Further, the Phase III trial is conclusive of the dose and clinical response as used in different stages of diseases and how the drug responds. The dose is also determined between the drug interaction with the co-administered drugs and food–drug interactions. Phase III is the detrimental stage when the drug could be prescribed with the necessary information.

The approved Ayurvedic drug/patent or proprietary medicines from outside India require testing of Indian population for further evidence of efficacy and safety prior to conducting Phase III studies in Indian participants. Before the Phase III clinical trial, the requisite permission is required from the DCGI, along with the application for the multicentric trial.

7.13.7 Phase IV: Post-Marketing Trials

The trial is conducted by the independent agency usually appointed by the Central Licensing Approving Authority (CLA) to understand the establishment of the drugs as useful of the drug-drug interaction(s), safety studies, dose-response mortality/morbidity studies, and pharmacodynamics studies and more accurately demonstrated by efficacy and the approved indication(s).

7.14 SCHEDULE E-1 OF AYURVEDIC DRUGS

7.14.1 Drugs of Plant Origin

Drugs that originate from herbal origin are listed as follows.

Bhallataka: Semecarpus anacardium Linn. F.
Ahipena: Papaver somniferum Linn.
Bhanga (except seeds): Cannabis sativa Linn.
Danti: Baliospermum montanum Mall. Arg.
Parasilka Yavani: Hyocyamus inibar Linn.
Jaipala (seed): Croton tiglium Linn.
Karaveera: Nerium indicum Mill.
Langali: Gloriosa superba Linn.
Parasilka: Yavani-Hyocyamus inibar Linn.
Vatsanabha/Shringivisha: Acontium chasmanthum Stapfex Holm
Vishmushti: Strychnox nuxvolnica Linn.

7.14.2 Drug of Animal Origin

Sarpa Visha: Snake poison.

7.14.3 Drugs of Mineral Origin

Gauripashna: Arsenic.
Hartala: Arseno sulfide.
Manahashila: Arseno sulfide.
Parada: Mercury.
Rasa Karpura: Hydragyri subchloridum.
Tuttha: Copper Sulfate.
Hingula: Cinnabar.

7.15 SCHEDULE E-1 OF UNANI SYSTEM

7.15.1 Drugs of Vegetable Origin

Afiyun (except seed): *Papaver somniferum* Linn.
Bazrul-banj: Hyoscyamus niger Linn.
Bish: Aconitum chasmanthum Stapfex Holmes.
Bhang (except seed): *Cannabis sativa* Linn.
Charas (resin): *Cannabis sativa* Linn. (except seed).
Dhatura seeds: Datura metal Linn.
Kuchla: Strychnos nuxvomica Linn.
Shokran: Conium maculatum Linn.

7.15.2 Drugs of Animal Origin

Sanp (head): Snake (head)
Telni makkhi: Mylabaris cichorii Linn, *Mylaboris pustulata* Thumb.

7.15.3 Drugs of Mineral Origin

Darachikna: Hydrargyri perchloridum
Hira: Diamond
Ras: Kapoor-Hydrargyri
Calomel: Subchloridum
Shingruf: Hydrargyri bisulphuratum
Zangar: Cupri subaccetas
Sammul-Far-Abyaz, Asfar, Aswad, and Ahmar: White, yellow, black, and red arsenic, respectively
Tootiya: Copper sulfate
Para: Hydrargyrum
Hartal: Arsenic trisulphide (yellow)

7.16 SIDDHA SYSTEM

Abini (except seed): *Papaver somniferum Linn.*
Alari: Nerium indicum Mill.
Attru Thumatti: Citrullus colocynthis Scharad.
Umathai: Datura stramonium Linn.
Etti: Strychnos nuxvomica Linn.
Ganja (except seed): Cannabis sativa Linn.
Kalappaik Kizhangu: Gloriosa superba Linn.
Kodikkalli (exempted for external use): *Euphorbia tituqalli* Linn.
Chadurakkalli (exempted for external use): *Euphorbia antiquorium* Linn.
Kattu Thumatti: Cucumis trigonus Roxb.
Kunri (except root): Arbus precatorius Linn.
Cheramkottai: Semicarpits anacardium Linn.
Thillai: Enoecaria agallocha Linn.
Nabi: Aconitum ferox Wall.

Nervalam: Croton tiglium Linn.
Pugaielai: Nicotiana tobacum Linn.
Mancevikkalli (exempted for external use): *Euphorbia* sp.

7.17 PROVISIONS OF AYURVEDIC DRUG APPROVAL

Any traditional Ayurvedic medication that satisfies the same textual indications may be admitted straight into Phase III/IV trials. Any licensed Ayurvedic medication having novel uses, including patent or proprietary medications may directly enter to the Phase II trial. Any ASU drug mentioned in Schedule E-I ingredients or patent or proprietary medicines with Phase I trials may be conducted as appropriate.

7.18 ETHNOGRAPHIC DESIGN FOR FOLK CLAIMS

The claims made by traditional healers hold significant importance within their local socio-economic and cultural contexts. Ethnographic design serves as a research methodology to validate these claims by providing a comprehensive understanding of traditional healing practices and their impact on the health and well-being of the community. The claim may not have the scientific data or documents but have originated from strong belief and social background. Thus, it is the challenge to establish the claim scientifically and accept validation methods. Despite these challenges, clinical trials for folk medicine are essential to establish the safety and efficacy of these practices, protect public health, and promote the integration of traditional and modern healthcare systems where appropriate. Addressing these challenges often requires a collaborative, culturally sensitive, and interdisciplinary approach involving healthcare professionals, researchers, and the local community.

Given these complexities, researchers conducting trials involving folk medicine often adopt alternative study designs. These may include controlled trials with active comparators, non-inferiority trials, or observational studies that focus on safety and efficacy within the context of holistic healing.

Ultimately, the decision on whether to include a placebo group in a clinical trial involving folk medicine depends on the specific research goals, cultural considerations, and ethical principles. Researchers must carefully weigh these factors to design studies that respect the unique nature of traditional healing practices while ensuring scientific rigor and patient safety. Second, most folk medicines are composed of multiple combinations of the plant drugs or poly-formulations, so it is difficult while selecting the placebo control to prepare the same smell, color, or texture of the combination formulations. Third, designing placebo is not possible for the process of therapy or procedural intervention of *Panchakarma* therapy in Ayurveda and so on. The use of a placebo is practical approach to understand the effects after intervention as compared to the no-treatment group and generates evidence of effects of treatment.

Thus, control groups are used alternate to the placebo groups whereby the data generates only without intervention for the same group before treatment and after treatment.

7.19 BLIND ASSESSMENT

Blind assessment is one of the approved methods of conventional evaluation of therapeutic interventions, but the practically is complicated in Ayurvedic medicine clinical trials. In the Ayurvedic clinical trial, any predetermined formulation is not administered. Blind trials are not a feasible option for procedure-based therapies like *Jalauka* therapy, *Panchakarma* therapy, *Ksharasutra,* and *Agnikarma*; further, blind procedures would not be possible for multi-drug–associated treatment in Ayurveda as in conventional traditional medicine whereby the shape, size, taste, weight, and active agents are used as active and control formulations with similar appearance, taste, and weight.

7.20 COMPONENTS OF THE ASU DRUGS

7.20.1 Pratisthapana or Research Question/Hypothesis

The research undertaken for the project is used to understand the importance of the disease and the prevalence or the burden of the disease to common people. Sufficient information is to be addressed about the research gap of the existing disease. The hypothesis of the research shall include the novelty of the proposed innovations and avoid to repetition of the study in similar objectives which may be same geographical location, variables parameters, populations, etc. Also, the justification is required for the volunteer recruitment with special reference of the age, gender, or if any venerable population group is associated to the clinical trial.

7.20.2 Interventions (if Applicable)

The details of the investigational products(s) or projected of interventions—including route of administration, dosage, intervention periods, dosing regimen, or behavioral intervention methods with the justification—is required to be submitted.

Additionally, any clinical research as well as any pre-clinical data from the *in vitro* or *in vivo* investigations that have clinical value should be mentioned.

7.20.3 Potential Risks

Identification of the potential risks of the involved study or intervention to human participants from the relevant literature is to be established and background information on the study's and the interventions' dangers to human subjects should be given. Further, any risk involvement to the pre-clinical study must be provided.

7.20.4 *Panchakarma* Procedures

Designing a clinical trial for Panchakarma procedures requires a deep understanding of Ayurvedic principles, the specific disease under investigation, and the cultural context. Collaboration with Ayurvedic practitioners and experts in Ayurvedic medicine is often crucial to the success of such trials. Additionally, close consultation with ethical review boards is essential to ensure that the trial respects cultural traditions and ethical standards.

7.20.5 *Poorvakarma* Procedures

These are also known as preparatory procedures and are sub-classified as *Snehana* (internal/external use of oil/ghee), *Deepaniya* (stomachic) and *Pachaniya* (digestants) and *Swedana* (medicated sudation). Such products require many documents for clinical trial. The source of trial drugs and the preparation procedures of drugs are to be mentioned. The clinical information like route, time, and duration of administration is to be mentioned in the trial dossier.

7.20.6 *Pradhankarma* Procedures

In Ayurvedic, formulation is known as main procedures and sub-classified as *Vaman* and *Zrma* (therapeutic emesis), *Virechan Karma* (therapeutic purgation), *Anuvasan basti* (oil/unctuous enema), *Asthapana Basti* (decoction-based enema), and *Nasya Karma* (nasal administration of medications). Such products require information during the clinical trial. Similar to the Poorvakarma procedure, they also require the source of trial drug and the preparation procedures of drug(s). The clinical information like route, time, and duration of administration is to be mentioned in the trial dossier. Also, the selected vehicle—along with the justification with the end point procedures or *Samyak lakshan*—has to be mentioned.

7.20.7 Para-Surgical Procedures

In Ayurveda, the para-surgical procedure is well defined and sub-classified as *Ksharasutra, Agni karma*, and *Rakta Mokshana*.

7.20.7.1 *Ksharasutra*

The sources of the sutures or threads, description of the materials, methods and drugs preparation procedures, and other clinical data like route, time, and duration of administration are to be mentioned.

7.20.7.2 *Agni Karma*

The sources of the material, description of materials, details of methodology, time of administration, duration of the action, and pre- and post-operative procedures are to be mentioned.

7.20.7.3 *Rakta Mokshana*

The sources of the material, description of material, details methodology, time of administration, duration of the action, pre- and post-operative procedures, and Samyak lakshan have to be mentioned.

7.21 PROVISIONING OF GOOD MANUFACTURING PRACTICE

As per the notification of the government of India, "Schedule T" vide GSR No. 560(E) on 23 June 2000 under Rule 157 amended the good manufacturing practice (GMP) vide GSR No. 198(E), dated 7 March 2003 of the Drugs and Cosmetic Rules.

These rules are enforced for all ASU drug manufacturing units from 23 June 2000 and for existing units from 23 June 2002, respectively, and a grace period of two years is given to all ASU manufacturing units which have to certify from the GMP certification.

Schedule T described the factory premises and standard hygienic conditions for manufacturing process of ASU drugs and all SU units to maintain the compliance with GMP. According the GMP, part-I and part-II, ensured the following conditions to be maintained in the process or manufacturing of ASU drugs.

7.21.1 Raw Materials

During the manufacturing process, the quality control of the raw materials shall be maintained and should be free from any contamination.

7.21.2 Quality Control

Quality control is the integral part of the Schedule T. In-process quality control and end products quality control are to be maintained according to the standard, and end products shall be accepted as per quality required.

7.21.3 Manufacturing Process

The manufacturing process shall maintain the minimum standards and the in-processes quality controls to be monitored. The end products shall qualify by minimum standards.

7.21.4 Documentation

The processes of control, in-process control, and quality control are documented, and related material shall be kept for reference and inspection.

Registered Vaidyas, Siddhas, and Hakeems under CCIM Act 1970 are prepared or dispense the medicine for their patients, and not marketing such drugs in the market is exempted from the purview of good manufacturing practice (GMP).

7.22 STANDARDIZATION OF ASU DRUGS

The standardization of the drugs depends on the sources—whether it is a raw plant drug, plant extract, or Avaleha/Leham/Mofak/Ilgam (confectionary or semi-solid).

7.22.1 Raw Plant Drugs

Raw plant drugs are crude drugs and are used directly by Ayurveda. However, raw plant drugs have to pass quality control. The following data is required for the approval of the raw plant drugs as ASU drugs.

- The plant details including the pleas of collection, season, etc.
- The botanical of the plants, including the place of collection, part used, and botanical description.
- The details of adulteration, if any.
- The foreign matter, if any.

- The major organoleptic characteristics, including the odor, color, taste, texture, etc.
- The identifying characters including the macroscopic, microscopic, and powder microscopy.
- Moisture content in percentage basis of total weight of the plants and the quantitative value of loss on drying at 105°C.
- Extractive value of 10% aqueous extract.
- The total value of the ash content.
- Value of acid-insoluble ash
- Value of water-soluble extract
- Extractive value of alcohol-soluble ash.
- Volatile oil content, if any.
- TLC/HPTLC/HPLC/GLC spectrum or values.
- Active constituents including tannins/total alkaloids/resins, etc., present in the formulation.
- As per the ASU Pharmacopoeia method to test for heavy or hazardous metals such as lead, cadmium, mercury, and arsenic.
- Confirmation of the content of pesticide residue, organophosphorus pesticides, pyrethroids, and organo-chlorine pesticides within the limits, as per ASU Pharmacopoeia.
- Details of the microbial contamination, including about the Enterobacteriaceae total viable aerobic count.
- Details of the fungal count within the limits as per ASU Pharmacopoeia.
- Specific pathogen contents like the *Escherichia coli*, *Salmonella spp.*, *Staphylococcus aureus*, and *Pseudomonas aeruginosa* within the limits as per ASU Pharmacopoeia.
- Details about the content of aflatoxins (B-1, B-2 and G1, G2) within the limits, as per ASU Pharmacopoeia.
- Shelf life.

7.22.2 Standardization of Plant Extracts as Drugs

Plant extracts are also used in Ayurveda. The following data is required for the approval of the plant extract as ASU drugs.

- The plant details, including the place of collection, season, etc.
- The botanical of the plants, including the place of collection, part used, and botanical description.
- The major organoleptic characteristics, including the odor, color, taste, and texture.
- The total value of the ash content.
- Value of acid-insoluble ash.
- Value of water-soluble extract.
- Extractive value of alcohol-soluble ash.
- pH of 10% aqueous extract.
- Plant extractive values like bulk density and tap density.
- Content of volatile oil.
- Residual solvent present shall be reported on a percentage basis.

- Active constituents, including tannins/total alkaloids/resins, etc., present in the formulation.
- TLC/HPTLC/HPLC/GLC spectrum or values.
- As per the ASU Pharmacopoeia method to test for heavy or hazardous metals such as lead, cadmium, mercury, and arsenic.
- Confirmation of the content of pesticide residue, organophosphorus pesticides, pyrethroids, and organo-chlorine pesticides within the limits, as per ASU Pharmacopoeia.
- Details of the microbial contamination, including about the Enterobacteriaceae total viable aerobic count within the limits, as per ASU Pharmacopoeia.
- Specific pathogen contents like *Escherichia coli*, *Salmonella spp.*, *Staphylococcus aureus*, and *Pseudomonas aeruginosa* within the limits, as per ASU Pharmacopoeia.
- Details about the content of aflatoxins (B-1, B-2 and G1, G2) within the limits, as per ASU Pharmacopoeia.
- Shelf life.

7.22.3 Standardization of *Avaleha, Leham, Mofak, Ilgam* (Confectionary or Semi-Solid)

- The plant details, including the pleas of collection, season, etc.
- The botanical of the plants, including the place of collection, part used, and botanical description.
- The major organoleptic characteristics, including the odor, color, taste, texture, and consistency.
- The total value of the ash content.
- Value of acid-insoluble ash
- Value of water-soluble extract
- Extractive value of alcohol-soluble ash.
- pH of 10% aqueous extract.
- Moisture content in percentage basis of total weight of the plants and the quantitative value of loss on drying at 105° C.
- Acidity of the formulation.
- Specific gravity of the formulation at 25°C.
- Total content of solid.
- Total content of fat.
- Total content of the reducing sugar/non-reducing sugar.
- Total sugars content.
- Major constituents or the main ingredients.
- Assay of the major ingredients or major constituents of main ingredients shall be reported.
- Any spectrum of the TLC/HPTLC/LC-MS/HPLC.
- As per the ASU Pharmacopoeia method to test for heavy or hazardous metals such as lead, cadmium, mercury, and arsenic
- Confirmation of the content of pesticide residue, organophosphorus pesticides, pyrethroids, and organo-chlorine pesticides within the limits, as per ASU Pharmacopoeia.

- Details of the microbial contamination, including about the Enterobacteriaceae total viable aerobic count within the limits, as per ASU Pharmacopoeia.
- Specific pathogen contents like *Escherichia coli*, *Salmonella spp.*, *Staphylococcus aureus*, and *Pseudomonas aeruginosa* within the limits, as per ASU Pharmacopoeia.
- Details about the content of aflatoxins (B-1, B-2 and G1, G2) within the limits, as per ASU Pharmacopoeia.
- Shelf life.

7.22.4 Standardization of Chewing Candy

- The plant details, including the method of collection, season, etc.
- The botanical details of the plants including the place of collection, part used, and botanical description.
- The major organoleptic characteristics, including the odor, color, taste, and texture.
- The identifying characters, including the macroscopic, microscopic, and powder microscopy.
- Moisture content in percentage basis of total weight of the plants and the quantitative value of loss on drying at 105^0 C.
- The candy characterization, including friability, hardness, and uniformity of weight (single dose shall not exceeds more than 5 mg).
- Disintegration time within the limit and shall be disintegrated in more than 30 minutes in phosphate buffer pH 6.4.
- pH of the candy.
- The total value of the ash content.
- Value of acid-insoluble ash
- Value of water-soluble extract
- Extractive value of alcohol-soluble ash.
- Salt content.
- Total sugars content like reducing sugars, non-reducing sugars, etc.
- As per the ASU Pharmacopoeia method to test for heavy or hazardous metals such as lead, cadmium, mercury, and arsenic
- Details of the microbial contamination, including about the Enterobacteriaceae total viable aerobic count within the limits, as per ASU Pharmacopoeia.
- Details of count of bacterial contamination.
- Specific pathogen contents like *Salmonella spp.* and *Staphylococcus aureus* within the limits, as per ASU Pharmacopoeia.
- Details about the content of Aflatoxins (B-1, B-2 and G1, G2) within the limits, as per ASU Pharmacopoeia.
- Shelf life.

7.22.5 Standardization of *Churna/Choornam* or Fine Powder/*Kvatha Churna/Kundir Choornam* (Coarse Powder for Decoction)

- The description.
- The major organoleptic characteristics, including the odor, color, taste, and texture.
- The foreign materials or others.
- Powder microscopy for identifications.

- Particle size distribution and permitted for 80–100 mesh for Cuma; 40–60 mesh for Kvatha Cuma.
- Quantitative value of loss on drying at 105° C.
- The total value of the ash content.
- Value of acid-insoluble ash.
- Value of water-soluble extract.
- Extractive value of alcohol-soluble ash.
- pH (5% aqueous extract).
- Bulk density and tap density of the powder.
- Any spectrum of the TLC/HPTLC/LC-MS/HPLC.
- Assay for constituents group of compounds or specific ingredients.
- As per the ASU Pharmacopoeia method to test for heavy or hazardous metals such as lead, cadmium, mercury, and arsenic.
- Confirmation of the content of pesticide residue, organophosphorus pesticides, pyrethroids, or organo-chlorine pesticides within the limits, as per ASU Pharmacopoeia.
- Details of the microbial contamination, including about the Enterobacteriaceae total viable aerobic count within the limits, as per ASU Pharmacopoeia.
- Total fungal counts, as per specification ASU Pharmacopoeia.
- Details about the content of Aflatoxins (B-1, B-2 and G1, G2) within the limits, as per ASU Pharmacopoeia.
- Shelf life.

7.22.6 Standardization of *Lepa, Malhara, Kalimbu, Paisai*, Medicated Wax, Poultice, or Cream

- The description.
- The major characteristics, including the odor, color, taste, etc.
- The viscosity if the material is free flowing.
- Rancidity test.
- Microscopy.
- Spreadibility.
- Powder microscopy for identification.
- Acidity of the formulation.
- Particle size (if powdered drugs incorporated) of the ingredients.
- Any spectrum of the TLC/HPTLC/LC-MS/HPLC.
- Assay of the drugs, if any is required.
- Uniformity of content and total fat content.
- Thermal stability.
- pH.
- Quantitative value of loss on drying at 105° C.
- As per the ASU Pharmacopoeia method to test for heavy or hazardous metals such as lead, cadmium, mercury, and arsenic.
- Confirmation of the content of pesticide residue, organophosphorus pesticides, pyrethroids, and organo-chlorine pesticides within the limits, as per ASU Pharmacopoeia.
- Details of the microbial contamination, including about the Enterobacteriaceae total viable aerobic count within the limits, as per ASU Pharmacopoeia.

- Specific pathogen contents, like *Salmonella spp*. and *Staphylococcus aureus*, within the limits, as per ASU Pharmacopoeia.
- Total fungal counts, as per specification ASU Pharmacopoeia.
- Details about the content of Aflatoxins (B-1, B-2 and G1, G2) within the limits, as per ASU Pharmacopoeia.
- Shelf life.

7.22.7 Standardization of *Pisthi/Bhasma/Sidur* (Processed Fine Powder)/*Parpam*

- The description of the formulations.
- The major characteristics, including the odor, color, taste, etc.
- Quantitative value of loss on drying at 105^0 C.
- pH (1% aqueous extract).
- The total value of the ash content.
- Value of acid-insoluble ash.
- Value of water-soluble extract.
- Extractive value of alcohol-soluble ash.
- Sulfated ash value.
- Particle size.
- Test for heavy or toxic metals—lead, cadmium, mercury, and arsenic—as per ASU Pharmacopoeia.
- Assay for other elements.
- Fingerprinting of XPS/XRD/SEM/EDX/AFM IR/XRF (as per requirements).
- Other ASU properties like *Nishchandrica* (lusterless), *Rekha pumatva* (fine enough to enter within lines of finger), *Varitara* (floats on water), *Nirthoom* (smokeless), *Apumar Bhav* (irreversible), and *Niswadu* (tasteless),
- Shelf life.

7.22.8 Standardization of *Sharkara Sikta/Siruthugalgal* (Granules)

- The description of the formulations.
- The major characteristics, including the odor, color, taste, etc.
- Microscopic property.
- Bulk density and tap density of the powder; other physical properties like compressibility, ash content.
- The total value of the ash content.
- Value of acid-insoluble ash.
- Value of water-soluble extract.
- Extractive value of alcohol-soluble ash.
- Flow property.
- pH.
- Sugars content like total sugars, reducing sugars, and non-reducing sugars.
- Any spectrum of the TLC/HPTLC/LC-MS/HPLC.
- As per the ASU Pharmacopoeia method to test for heavy or hazardous metals such as lead, cadmium, mercury, and arsenic.
- Specific pathogen contents, like *Salmonella spp*. and *Staphylococcus aureus*, within the limits, as per ASU Pharmacopoeia.

- Confirmation of the content of pesticide residue, organophosphorus pesticides, pyrethroids, and organo-chlorine pesticides within the limits, as per ASU Pharmacopoeia.
- Total viable aerobic count within the limits, as per ASU Pharmacopoeia.
- Total fungal counts, as per specification in the ASU Pharmacopoeia.
- Specific pathogen contents, like *Salmonella spp*. and *Staphylococcus aureus*, within the limits, as per ASU Pharmacopoeia.
- Shelf life.
- Specific aflatoxins (limits as per ASU Pharmacopoeia) (B1, B2 and G,1 G2)

7.22.9 Standardization of *Sharbat/Manappagu* (Syrup)

- The description.
- The major characteristics, including the odor, color, and taste.
- The product viscosity, pH, and total solids.
- Sugars content like total sugars, reducing sugars, and non-reducing sugars.
- Specific gravity at 25°C.
- Any spectrum of the TLC/HPTLC/LC-MS/HPLC.
- As per the ASU Pharmacopoeia method to test for heavy or hazardous metals such as lead, cadmium, mercury, and arsenic.
- Confirmation of the content of pesticide residue, organophosphorus pesticides, pyrethroids, and organo-chlorine pesticides within the limits, as per ASU Pharmacopoeia.
- Details of the microbial contamination, including about the Enterobacteriaceae total viable aerobic count within the limits, as per ASU Pharmacopoeia.
- Total fungal counts as per specification ASU Pharmacopoeia.
- Specific pathogen contents, like *Salmonella spp*. and *Staphylococcus aureus*, within the limits, as per ASU Pharmacopoeia.
- Specific aflatoxins (limits as per ASU Pharmacopoeia) (B1, B2 and G1, G2).
- Shelf life.

7.22.10 Standardization of Tailas, Thylam, Nei, or Ghiritas (Medicated Oils and Ghee)

- The description.
- The major characteristics, including the odor, color, and taste.
- The product rancidity values.
- Specific gravity at 25ºC.
- Quantitative value of loss on drying at 105ºC.
- Rancidity values.
- pH value.
- Congealing point, refractive index, and viscosity.
- Iodine value, saponification value, unsaponifiable matter (percentage of).
- Quantitative values like acid value, iodine value, and saponification value.
- Peroxide value.
- Free fatty acids and total fatty acid content.
- Free from mineral oil.

- Any spectrum of the TLC/HPTLC/LC-MS/HPLC.
- As per the ASU Pharmacopoeia method to test for heavy or hazardous metals such as lead, cadmium, mercury, and arsenic.
- Confirmation of the content of pesticide residue, organophosphorus pesticides, pyrethroids, and organo-chlorine pesticides within the limits as per ASU Pharmacopoeia.
- Details of the microbial contamination, including about the Enterobacteriaceae total viable aerobic count within the limits as per ASU Pharmacopoeia.
- Specific pathogen contents like Salmonella spp., Staphylococcus aureus, within the limits as per ASU Pharmacopoeia.
- Specific Aflatoxins (Limits as per ASU Pharmacopoeia) (B1, B2, G1 G2).
- Shelf life.

7.22.11 Standardization of Vati, *Gutika, Kuligai, Mathirai*, or Vadagam (Tablet or Pills)

- The description.
- The major characteristics, including the odor, color, taste.
- Microscopic characters if any vegetable parts used as ingredient.
- pH of 5% aqueous extract.
- Any spectrum of the TLC/HPTLC/LC-MS/HPLC.
- Quantitative value of loss on drying at 105°C.
- Tablet characteristics like friability, hardness and uniformity of weight.
- Disintegration time usually within the limit of 35 minutes except Guggulu.
- The ash content including the three extractive values: alcohol-soluble, water-soluble, and acid-insoluble ash.
- Alcohol, water—soluble extractive.
- Total sugar content: Reducing sugar/Non-reducing sugar.
- Assay of the products' active constituents of major ingredients.
- As per the ASU Pharmacopoeia method to test for heavy or hazardous metals such as lead, cadmium, mercury, and arsenic.
- Confirmation of the content of pesticide residue, organophosphorus pesticides, pyrethroids, organo-chlorine pesticides within the limits, as per ASU Pharmacopoeia.
- Details of the microbial contamination, including about the Enterobacteriaceae total viable aerobic count within the limits, as per ASU Pharmacopoeia.
- Specific pathogen contents like *Salmonella spp*. and *Staphylococcus aureus*, within the limits as per ASU Pharmacopoeia.
- Specific aflatoxins (limits as per ASU Pharmacopoeia) (B1, B2 and G1, G2).
- Shelf life.

7.23 LICENSE APPROVAL AND THE CONDITIONS

7.23.1 Section 3(a) of Drugs and Cosmetics Act, 1940: Ayurveda, Siddha, and Unani Medicines

The formulations are delineated in the authorized literature of Ayurvedic, Siddha, and Unani Tibb systems of medicine, as well as in the first schedules, which prescribe their internal or external application for the purposes of diagnosing, treating, alleviating, or preventing diseases.

7.23.2 Section 3(h) of Drugs and Cosmetics Act, 1940: Patent or Proprietary Medicine

The formulations included in the section 3(a) of Drugs and Cosmetics Act (1940) of first schedules of the Ayurvedic, Siddha, and Unani Tibb system are authoritative, but excludes medications that are given intravenously, as stated in the Drugs and section 3(h) of Drugs and Cosmetics Act (1940) gives direction for the parenteral application of Ayurvedic, Siddha, and Unani Tibb systems.

Further, Ayurvedic, Siddha, and Unani Tibb systems admit the positive health promoter, balya, poshak, muqawi, and unavupourtkal formulations which are under the section 3(a) of Drugs and Cosmetics Act (1940) as schedule 1 of the act used for promotional and preventive health.

The formulations used in the mouth, skin, hair, and body care are known as Azhagh-sadhan or the formulation known as *Saundarya Prasadak (Husane afza)* and is referenced in the First Schedule of the Drugs and Cosmetics Act.

The use of medicinal plant extracts, especially in traditional medicine systems like Ayurveda, is based on centuries of accumulated knowledge and experience. However, the quality, safety, and regulatory aspects of these extracts are subject to modern standards and regulations to ensure their suitability for therapeutic use. In India, the Ayurvedic Pharmacopoeia and related regulations provide guidance on the use of medicinal plant extracts in traditional medicine.

7.23.3 Conditions for the Issuing of a License

ASU drugs are manufactured exclusively to the mentioned formulae in the accepted literature. The scheduled book, as per the Drugs and Cosmetics Act (1940), is approved in the literature for ASU drugs. Other than any drugs is not accepted. The regulatory and quality standards for ASU drugs are in place to strike a balance between preserving traditional knowledge and ensuring that products are safe, effective, and consistent. These standards help maintain the integrity and authenticity of traditional medicine systems while addressing modern quality and safety requirements.
The conditions are described as follows.

Rules	Suggestions
Ayurveda, Siddha and Unani drugs, referred in section 3(a) of Drugs and Cosmetics Act (1940) clause of 158.B	Not required any study further proof of effectiveness or safety.
Ayurveda, Siddha and Unani drugs, referred in section 3(a) Drugs and Cosmetics Act (1940) but the change in dosage form	This category exempted the safety study and the proof of effectiveness, but the indication, experiences, and evidence of effectiveness related the literature are required.
Ayurveda, Siddha and Unani drugs, referred in section 3(a) Drugs and Cosmetics Act, Ayurveda, Siddha 1940 but referred in 3(a) to be used for new indication	This category exempted the safety study and the proof of effectiveness, but the indication, experiences and evidence of effectiveness related the literature are required.

Rules	Suggestions
Patent or proprietary medicine	There is no scope of the alternation of the ingredient and shall be according the text. This category exempted the safety study, but the proof of effectiveness is required or the clinical trial is required to establish the claim, as per the protocol for Ayurveda and Siddha and Unani drugs. Individual ingredients details are required, as per the established in literature.
Ayurveda, Siddha, and Unani drugs with any of the ingredients of Schedule E (1) of the Drugs and Cosmetics Act (1940)	The ingredient is listed in the schedule E-1, indications are already established in the literature, and additional safety studies are required. Further clinical trial to establish the literature and published literature support is required.

The licensing procedures for the Indian System of Medicine, which includes traditional systems such as Ayurveda, Siddha, and Unani, are governed by regulatory authorities in India. The primary regulatory body overseeing these systems is the Ministry of AYUSH (Ayurveda, Yoga & Naturopathy, Unani, Siddha, and Homoeopathy).

Here's an overview of the licensing procedures for practitioners and institutions in the Indian System of Medicine:

Educational Qualifications: Practitioners need to have the required educational qualifications from recognized institutions. The qualifications typically include undergraduate and postgraduate degrees in Ayurveda, Unani, Siddha, or other traditional systems.

Council Registration: Practitioners are required to register with the respective statutory councils for each system. For example, the Central Council of Indian Medicine (CCIM) regulates Ayurveda and Unani, while the Central Council of Siddha (CCS) oversees Siddha practitioners.

State Medical Boards: Each state in India may have its own State Medical Board or Council for the registration and licensing of practitioners. Practitioners often need to be registered with the state board to practice within that state.

Licensing Authorities: Licensing authorities may vary depending on the specific system of medicine. For Ayurveda, Unani, and Siddha, the respective councils mentioned above play a key role in licensing and regulating practitioners.

Institutional Licensing: Institutions offering education and training in the Indian System of Medicine need to adhere to guidelines set by the respective councils. These institutions are typically required to obtain recognition and approval from the appropriate regulatory bodies.

Quality Control and Good: Manufacturing Practices (GMP): Manufacturing units involved in the production of traditional medicines are required to adhere to quality control standards and Good Manufacturing Practices (GMP). The quality of medicines is crucial to ensure safety and efficacy.

Research and Development (R&D) Licensing: Entities involved in research and development in the field of traditional medicine may require specific approvals and licensing for their activities.

7.24 *AUSHADH GHANA* OR MEDICINAL PLANT EXTRACT

Further, the license requires for the extract of medicinal plants or Aushadh Ghana in different conditions or the category. License required for different formulation types for extract medicinal plants as as follows:

Aqueous: This category exempted the safety study and the proof of effectiveness, but there is no scope alternation of the ingredients indications from the literature as mentioned.

Aqueous formulation with the new indication other than mentioned in 1st schedule books of the Drugs and Cosmetics Act (1940): This category exempted the safety study, but requires proof of effectiveness, i.e., clinical trial. There is no scope alternation of the ingredient, but the new indications claim is endorsed or permitted in this category.

Hydro-Alcohol: This category requires the safety study and requires proof of effectiveness, i.e., clinical trial. There is no scope alternation of the ingredient and indications.

Hydro-alcohol extract used for used in the new indication, i.e., other claims do not mention in the 1st schedule books of Drugs and Cosmetics Act (1940): This category requires the safety study and requires proof of effectiveness, i.e., clinical trial. There is no scope alternation of the ingredient and indications.

Any other than hydro/hydro-alcohol: This category requires the safety studies like acute, chronic, mutagenicity, and teratogenicity, and it requires proof of effectiveness, i.e., clinical trial.

BIBLIOGRAPHY

1. Safety/Toxicity Evaluation of the Ayurvedic Formulations Central Council for Research in Ayurvedic Sciences, Ministry of AYUSH, government of India, New Delhi, J. L. N. B. C. A. H. Anusandhan Bhavan, 61–65, Institutional Area, Opp. D-Block, Janakpuri, New Delhi—110 058, 2018.
2. General Guideline for Drug Development of Ayurvedic Formulations Central Council for Research in Ayurvedic Sciences, Ministry of AYUSH, Government of India, New Delhi, J. L. N. B. C. A. H. Anusandhan Bhavan, 61–65, Institutional Area, Opp. D-Block, Janakpuri, New Delhi—110 058, 2018.
3. General Guideline for Clinical Trials for Clinical Evaluation Ayurvedic Intervention Central Council for Research in Ayurvedic Sciences, Ministry of AYUSH, Government of India, New Delhi, J. L. N. B. C. A. H. Anusandhan Bhavan, 61–65, Institutional Area, Opp. D-Block, Janakpuri, New Delhi—110 058, 2018.
4. Gazette Notification, Ministry of Health Family Welfare, Department of Ayurveda, Yoga, Naturopathy, Unnani, Siddha and Homeopathy. Published ed in September 2014.
5. Shrivastava SR, Shrivastava PS, Ramasamy JJ. Mainstreaming of Ayurveda, Yoga, Naturopathy, Unnani, Siddha and Homeopathy with the Health Care System in India. *J Tradit Complement Med.* 2015; Jan 7;5(2): 116–118.
6. Nooreen Z*, Rai VK*, Yadav NP. Phytopharmaceuticals: A New Class of Drug in India. *Ann Phytomed.* 2018; 7(1): 27–37.

8 Regulatory Approval Processes of Phytopharmaceuticals, Herbal Drugs, or Isolated Compounds from Plants or New Claims Other than ASU Drugs

8.1 INTRODUCTION

As per the Drugs and Cosmetics Act of 1940, phytopharmaceutical drugs should have a minimum of four bioactive compounds or phytochemical compounds. Phytopharmaceuticals are both qualitatively and quantitatively extracted medicinal plants, or the parts can be used internally or externally by humans or animals for the treatment, diagnosis, mitigation, or protection of any illness or disorder; however, parental use is not permitted in all cases.

The rigorous regulatory regulation of phytopharmaceuticals is enacted and accepted by the scientific community after thorough scientific validation, but many standpoints may arise.

First, small and medium industries cannot participate in the development of pharmaceuticals and the delivery of phytopharmaceuticals is not economical. Second, no process of approval of phytopharmaceuticals consisting of regularized well-defined guidelines is available. Third, as per regulations, the process goes to the clinical trial to establish efficacy, and it is therefore a costly affair. Fourth, the intellectual property of phytopharmaceuticals is not easy because the costly affairs of the international patent process and the regulation of the Biological Diversity Act 2002 also affect phytopharmaceuticals.

8.2 LEGITIMATE PHYTOPHARMACEUTICAL REGULATIONS

Phytopharmaceuticals are under the regulation of Chapter IV of Drugs and Cosmetics Act, 1940, and similar to items made from synthetic drugs. The current regulations for the regulation of ASU pharmaceuticals are outlined in Chapter IVA

DOI: 10.1201/9781003398431-8

of the Pharmaceuticals and Cosmetics Act and Rules. Therefore, the aforementioned proposed regulations do not currently have any influence on the existing regulatory framework for ASU drugs. As per Rule 2 (eb) of the Drugs and Cosmetics Rules, 1945, section 122E, a phytopharmaceutical could not be considered as a drug until the safety data is established. Further, section 21 of Drugs and Cosmetics Rules, 1945, states that the proposed claims could be considered as a new ASU drug for treatment or diagnosis of disease only when the formulas are described in authoritative books of Ayurvedic, Siddha, and Unani Tibb systems of the first schedule. According to the gazette notification G.S.R.702 € dated 24 October 2013, "phytopharmaceutical drug" relates to any material processed from the materials derived from plants or parts thereof, including the extracts or fractions used for treatment, diagnosis, or mitigation or prevention of disease to animals or humans. But phytopharmaceuticals cannot be used parenterally. Schedule E drugs and hydroalcoholic extracts or other solvent extracts are mandatory for toxicity studies, while under section 22 of the Food Safety and Standards Act of India (FSSAI), a few of 180 herbs are approved which are known as "phytopharmaceutical drugs". Pharmaceutical ingredients or phytopharmaceutical drugs in the USFDA system is considered as new drug application (NDA), and meeting of specific requirements are required for approval of an investigational new drug application (INDA). Central Drugs Standards Control Organisation (CDSCO) is responsible for regulating phytopharmaceuticals while ASU drug approval is under the purview of the State Licensing Authority. This gazette notification is directed to the regulatory requirement of scientific data on the quality, safety, and efficacy of phytopharmaceuticals which is quite similar to new drug approval. The provision of a clinical trial as mentioned in Schedule Y of the Drugs and Cosmetics Act is enforced as it applies to phytopharmaceuticals.

While Appendix I B of said Act describes the data required for the clinical trial for importing or manufacturing of phytopharmaceutical drugs in the country. According to the gazette notification G.S.R.702 (E), all mandatory requirements for phytopharmaceutical drugs are required to be fulfilled with new drug approval like pre-clinical research, validating clinical trials, and information on the safety of drugs and pharmacology.

8.3 REGULATION OF PHYTOPHARMACEUTICALS: A NEW DIRECTIVE

Phytopharmaceuticals have a wider scope than Ayurvedic, Unani, or Siddha medicine because they can be used phytopharmaceuticals other than what is described in the literature of the first scheduled of AYUSH regulation. Further, phytopharmaceuticals could be used in combinations, or new herbs including herbals of foreign origin, in combination with vitamins and minerals. Further, traditional extraction methods could be avoided, and the extraction process follows advanced techniques which include accelerated solvent extraction, fractionation, freeze-drying, and supercritical extraction for modern formulation development.

The guideline strongly advocates the scientific rationality of the development of phytopharmaceuticals with comparison to any advanced country like the United States, Japan, or China. The guideline is also not dependent solely on the traditional knowledge alone as ASU medicine. Thus, phytopharmaceutical regulatory guidelines are based on more innovations and the development opportunities of new drugs from plant origins with the validation of science and which are encouraged to export to the advanced country. Further, the guideline is also established with the structure-activity relationship, and most modern medicines support this fundamental. Several traditional medicines (TM) like traditional Chinese medicine, Kampo medicine, Bhutanese medicines, Tibetan medicines, and ASU which are based on the traditional knowledge are established in the limited medicine texts. Though the TM medicine also validates their scientific claim, it is quite rational that many medicines could not explore scientifically and be documented as modern medicine from the ocean of Ayurveda drugs. CDSCO approved many phytopharmaceuticals like Guggulu tablets (for treatment of hypercholesterolemia), *Ginkgo biloba* tablets (to treat temporary loss of memory), and Silymarin capsules (to treat liver disorders). Also, USFDA approved a topical cream containing standardized green tea extract for the treatment of genital warts. As stated earlier, phytopharmaceuticals are the same as the category of the new chemical entity (NCE) and are mandatory to follow same guideline as synthetic compound-based pharmaceuticals are evaluated for safety (toxicology) and effectiveness through carefully carried out by human clinical trials.

8.4 STANDARDIZATION OF PHYTOPHARMACEUTICALS

As per Indian Pharmacopoeia, 2014, standardization of any extract means an extract having the biomarker or chemical/analytical marker within an acceptable tolerance. Standardization of the formulation may be accomplished by mixing one or more batches of extracts, modifying the extracts using authorized inert material, or both. Any phytopharmaceutical likely to establish as drugs requires an established minimum of four compounds identified as bioactive/phytochemical compounds. Tracing of bioactive/phytochemical compounds is required for maintaining batch-to-batch consistency, and the justification should be submitted by the applicant.

Primary standardization methods adopted are identifying a geo-climatically appropriate location, growth conditions, plant growth, harvesting time, collection time, drying, washing, and storage conditions. Further, the standardization of the plant's products also requires the proper guideline of the plant material's dealing with, garbling, shipping, crushing, and pulverization.

In the standardization process, many parameters are included like the authentication of plant material, uniform particle size, and presence of foreign matter, phytotoxins, radioactive contaminants, volatile matter, etc. Also, organoleptic characteristics—like macroscopy and microscopy, solvent residues, ash values, microbial count, pesticide residues, heavy metal residues, and mycotoxin residues—are to be evaluated.

Accordingly, in Schedule Y, Appendix I B, the clinical trial data is required for the production or import of a phytopharmaceutical product within the Nation.

This gazette notification also directed the requisite data on quality, safety, and efficacy, and it is required for marketing and is quite similar to the synthetic, chemical moieties, or new chemical moiety drugs. Human trials also require establishing the safety and efficacy of certain data, as per guidelines for new drugs. Phase I data is required to access the maximum tolerated dose and associated toxicities. Certain conditions studies may be abbreviated, modified, or relaxed according to the established or published literature or using the last five years without any toxicities.

8.5 MARKET APPROVAL

The Gazette notification by the government of India vide notification number GSR 826 (E), dated 30 October 2015. F. No. X. 11014/2/2012-DFQC GSR 918(E) amended the Drugs and Cosmetics Rules, 1945. This notification is a clear directive of approval procedure of ASU drugs, which as per Appendix IB are sub-divided into many categories describing the data requirement, clinical pharmacology, identification, authentication, etc. Any phytopharmaceutical drugs could be applied through Form CT-21 to the Central Licensing Authority, along with requisite data specified, accompanied with a fee as specified in the sixth schedule.

8.6 DETAILS OF APPLICATION OF APPROVAL

1. Applicant details, address, manufacturing premises, and phone number.
2. Name of the drug for which the license is sought.
3. Bioactive constituents and phytopharmaceuticals (qualitatively and quantitatively assessed).
4. Proposed indication of the treatment of diseases.
5. Complete application Form 44 with the authorized signatures with name and designation.
6. Treasury challan as per Schedule 6 and state the new drug's name, together with the account's correct name, bank authorization, etc.
7. Manufacturing license in Form 25/28, along with copy of Form 29 issued by the State Authority.
8. Source of bulk drugs or phytopharmaceuticals.
9. In-house and national biodiversity clearance is required.
10. If from other than in-house source, a copy of consent letter from manufacturer is required.

8.7 MANDATORY DATA REQUIREMENTS

1. An overview of the phytopharmaceutical drug together with the plant's botanical name of the plant which includes vernacular or scriptural name. Formulation details, along with the route of administration, dosages, therapeutic class, and the claims.
2. Available literature in terms of documented anciently or published anywhere to establish the plant product or phytopharmaceutical drug is a considered as a traditional medicine or as an ethnomedicine. Further, if any literature mentions the composition, dose, or method of usage, or proportion of the active ingredient and as per daily consumption and uses.
3. Any information or directives regarding the formulation's present use, adverse reactions of traditional or ethnomedicine, and contraindications.
4. Any scientific reports related to studies on pharmacology and safety that are necessary for the marketing approval of phytopharmaceutical products.
   a. Any known data available about the use of similar products as traditional medicine or ethnomedicine.

   b. Different use of the pharmaceuticals as traditional medicine or ethnomedicine.
5. Any contraindications or side effects are reported, or any adverse reactions reported of the phytopharmaceutical in the last three years.
6. Status of the present phytopharmaceutical drug, whether using or the details of product, quantum sold, manufacturer, effects of exposure on human population, and since when the product has been marketed.

8.8 PUBLISHED SAFETY DATA, IF ANY

1. Any published scientific reports (which includes pharmacological studies, human studies, epidemiological studies, clinical studies, etc.) which are relevant to the phytopharmaceutical drug.
   a. If the same or similar products are used in traditional or ethnomedicine.
   b. If the same or similar products are used in different forms in traditional or ethnomedicine.
2. Pharmacodynamic information (if available).
3. If any monograph (English translation) is available or written about, the same medicinal plant, herbal remedy, extract, or phytopharmaceutical.
4. Extracted Plant or fractionation identification, authentication, and source.
5. Plant taxonomical genus, species, family, taxonomic reference, and cultivar name (if applicable) are all required where the phytopharmaceuticals are collected.
6. Identity and authenticity of the plants along with the certificate for botanical identity by a qualified taxonomist. Also, the morphological and anatomical description, diagnostic features and a photograph of the plant or plant part are required.
7. Habitat and geographical distribution of the plant including the biodiversity importance—while the photochemical are isolated part of plant which is renewable or destructive process.
8. Collection time with seasonal variance.
9. Source of the plant with the geographical location and ideal season for collection.
10. The statement is required whether the phytopharmaceutical are produces any species which are:
    a. Endangered or threatened under the Endangered Species Act or the Convention on International Trade in Endangered species (CITES) of wild Fauna and Flora.
    b. Special protection under the Biological Diversity Act, 2002 (18 of 2003).
    c. Knowledge of genotypic, chemotypic, and ecotypic variability of species.
11. The flowing information is to be furnished of the grower or supplier (including names and addresses).
    a. Location of harvesting.
    b. A specific growth conditions.
    c. Plant growth at the time of harvesting.
    d. Time of harvesting.

   e. Collection, drying, washing, and storage conditions.
   f. Transportation, handling, garbling.
   g. Pulverizing and grinding of the plant material.
   h. Consistent particle size via sieving.
12. The flowing information are to be furnished as quality specifications.
   a. foreign matter if any
   b. total ash
   c. acid-insoluble ash
   d. pesticide residue

8.9 EXTRACTION PROCESS AND SUBSEQUENT FRACTIONATION AND PURIFICATION

1. Quality and standardization parameters or specifications with test methods for starting material.
2. Processing methodology.
3. Information about the solvent used, extraction values, solvent remaining tests or limits, physico-chemical tests, heavy metal impurities, microbial loads, chromatographic finger print profile with reference markers from phytochemistry, assay for active constituents or distinctive markers in the event that active constituents are unknown.
4. Characterization of final product after purification.
5. Bioactive constituent after purification.
6. Details of excipients, stabilizers, or preservative or diluents used.
7. Packaging details, including primary, secondary packaging, and purified and characterized final product, labeling and storage conditions.

8.10 FORMULATION DETAILS OF PHYTOPHARMACEUTICAL DRUGS

1. Formulation composition details including the purified phytopharmaceutical fractions in unit dose with defined markers and the other excipients, stabilizers, and other materials.
2. Identification of the phytopharmaceutical drug and the test methodology.
3. Chromatographic specifications of active and assay of an active ingredient or distinctive chemical signature.

8.11 MANUFACTURING PROCESSES AND ASSAY OF AN ACTIVE INGREDIENT OR DISTINCTIVE CHEMICAL SIGNATURE

1. The brief methodology of manufacturing of dosage form, including the environmental controls, in-process quality control and acceptance limit is to be mentioned.
2. Details of materials used in packaging, and all the description must be mentioned.

3. Quality specifications of the finished products, specification and it is necessary to record the dose form, chromatographic finger print profile, quality with phytochemical reference marker, and assay for the active ingredient or characteristic marker.

8.12 STABILITY DATA

1. Individual only phytopharmaceutical drug stability data at room temperature is required at 40 ±2°C, humidity at 75 ±5% RH (relative humidity), and tested for zero, one, two, three, and six months.
2. Formulation along with the packaging for final marketing of phytopharmaceutical drug the stability data is require at the room temperature at 40 ±2°C, humidity at 75 ± 5% RH, and tested for zero, one, two, three, and six months.

8.13 SAFETY AND PHARMACOLOGICAL DATA

1. Data on safety and pharmacological studies are required.
2. Pre-clinical toxicity and safety data.
3. By using two species of animals 28–90 days repeat dose oral toxicity data.
4. *In vitro* genotoxicity data and chromosomal aberration test as per Schedule Y.
5. Dermal toxicity tests for topical use.
6. Teratogenicity data applicable to the phytopharmaceutical drug is intended for use during pregnancy.

8.14 CLINICAL STUDIES

1. Clinical trials for phytopharmaceutical drugs are to be similar to new drug development or new chemical entities.
2. Phase I trial is required for estimation of the maximum of tolerated dose as related to toxicities. The clinical studies approval is required for conducting of the study.
3. Information about the protocols and the outcomes of the dose finding studies must be submitted for approval prior to performing the studies. The certain provision—abbreviated, modified may relaxed by DCGI.

8.15 CONFIRMATORY CLINICAL TRIALS

1. Safety and efficacy studies results are to be submitted for the phytopharmaceutical drug.
2. Phytopharmaceutical dosage for human clinical studies proposal for approval to generate and validate safety and efficacy data, as per applicable rules and guidelines.
3. The above investigations will maintain the formulation's quality control.

8.16 REGULATORY STATUS TO OTHER COUNTRIES

It's important for manufacturers and marketers of phytopharmaceutical drugs to be aware of and comply with the regulatory requirements of each country where their products are marketed. This includes meeting quality, safety, labeling, and health claim standards specific to the regulatory category under which the product is marketed. Regulatory compliance ensures the safety and efficacy of the product and helps prevent misrepresentation or unsafe marketing practice.

8.17 MARKETING INFORMATION

1. Details of leaflet and directions of the phytopharmaceuticals inserted to the packaging.
2. Draft of label and carton.

8.18 POST-MARKETING SURVEILLANCE (PMS)

1. After approval the drug for marketing, the manufacturer or market authority shall submit safety reports every six months.
2. After two years, the manufacturer or regulatory authority shall submit safety reports on an annual basis.

8.19 OTHER RELEVANT INFORMATION

Additional pertinent data that the applicant feels might be helpful in the application's scientific assessment.

BIBLIOGRAPHY

1. Nooreen, Z., Rai, V.K., Yadav, N.P. Phytopharmaceuticals: A new class of drug in India. *Annals of Phytomedicine* 2018; 7(1): 27–37.
2. AYUSH Guidelines, G.C.P. Good clinical practice guidelines for clinical trials of ASU medicine. 2013. Website: www.indianmedicine.nic.in/writereaddata/linkimages/5110899178-Final%20Book%2028–03–13.
3. Bhatt, A. Phytopharmaceuticals: A new drug class regulated in India. *Perspectives in Clinical Research*, 2016; 7: 59–61.
4. Gazette, N. Ministry of Health and Family Welfare Gazette Notification G.S.R. 918(E). Website: www.cdsco.nic.in/writereaddata/GSR%20918-E-dated-30-11 2015.pdf. Last accessed on 15 September 2020.
5. Vaidya AD, Devasagayam TP. Current status of herbal drugs in India: an overview. *J Clin Biochem Nutr*. 2007 Jul;41(1):1–11. doi: 10.3164/jcbn.2007001. PMID: 18392106; PMCID: PMC2274994.

9 Safety and Toxicity of Ayurveda, Siddha, and Unani (ASU) Drugs

9.1 INTRODUCTION

Each drug's classification in Ayurvedic texts is based on its distinctive properties and its pharmacological effects in the human system, including the physiological response, clinical efficacy, and therapy of the condition. It is necessary to locate and eliminate the component of Ayurvedic medicine that causes toxicity. According to Ayurveda, a person's body must be purified of toxins before their structural integrity can be restored. This is done through dietary adjustments, carefully selected natural remedies, and religious practices like yoga, meditation, and pranayama. About 600 herbal medications listed in Ayurvedic scriptures are categorized based on their characteristics and distinctive activities. The majority of herbs are categorized based on their biological functions, medicinal effects, and physiological properties. According to the Ayurvedic philosophy, each plant influences physiological balance and eliminates the disease-causing physiological imbalances that it causes.

Each medicine is categorized according to its distinctive qualities and pharmacological effects on the human body in Ayurvedic texts. The characteristics of herbal medications include physiological reaction, clinical effectiveness, and disorder treatment. The idea behind Ayurvedic medicine is that any abnormality of the human system needs to be found and treated. Toxins and ama must be removed from the human body. In Ayurveda, the treatment for mineral reinforcement is referred as "bhasmas". The conventional method for making "bhasmas" uses particular physical–chemical procedures designed to oxidize the minerals. According to popular belief, the production of "bhasmas" heavy metals is decontaminated and free of dangerous toxins. The majority of elements should be viewed as advantageous in their soluble forms.

9.2 CLASSIFICATION OF ASU MEDICATIONS BASED ON SAFETY

Safety is ASU medicines' top priority, and safety categories are divided into the following groups based on time periods.

- Category 1: Safety that has been long established by the user.
- Category 2: Under particular use circumstances, safety is covered by established documentation.
- Category 3: Uncertain safety, such that safety data must be established—in this class, safety data are occasionally the same as for any new chemical.

DOI: 10.1201/9781003398431-9

In accordance with the Ayurvedic pharmaceuticals' safety classification, assessing safety studies are needed before releasing herbal medications on the market, according to the Gazette's notification No. F.28–10/45-H(I), issued 21 December 1945, and its amendment No. GSR 602(E), dated 19 July 2010. No. K.11020/02/2010-DCC (AYUSH) classifies ASU medications based on their safety.

9.2.1 In Accordance with Section 3(a) of the Ayurvedic, Siddha, and Unani Medicines

As stated in the first schedule, ASU medications are produced in accordance with the authorized texts of the Siddha, Ayurvedic, and Unani Tibb systems. They are used for disease diagnosis, treatment, mitigation, or prevention in humans and animals.

9.2.2 Patented or Proprietary Ayurvedic, Siddha, and Unani Medicines Under Section 3(h)

The following medications are listed as being covered by section 3(h):

1. According to the first schedule, the ASU medications' ingredients are listed in authoritative books of the Ayurvedic, Siddha, or Unani Tibb system of medicine, but the formulation and mode of administration are not described in those publications, as is the case with section 3(a)'s reference to authoritative literature.
2. Ingredients of positive health promoters or formulations known as *Balya* or *Poshak* that are included in books of the First Schedule of the Drugs and Cosmetics Act and advised for use as positive health promoters.
3. *Saundarya Prasadak*, *Azhagh-sadhana*, or *Husane afza* formulations, which are named for ingredients used for oral, skin, body, or hair care and recommended for use as a positive health promoter, are listed in books of the First Schedule of the Drugs and Cosmetics Act, 1940. Ingredients or plants are used to make therapeutic plant extracts, either dry or wet, such as aqueous or hydro-alcohol extracts, or they are referred to as *Aushadh Ghana* in the act's first schedule volumes.
4. Ayurvedic, Siddha, and Unani medications are listed in 158-B as referred to section 3(a): The substance, indications of use, and no alternations are made. In these situations, neither a safety study nor any effectiveness evidence is necessary. However, these examples from the published literature are necessary to comprehend the efficacy.
5. Without any modification to the dose form of Ayurvedic, Siddha, or Unani medicines as per section 3(a) of the Drugs and Cosmetics Act of 1940, and Similar to what was stated before, the ingredients and usage instructions are exactly as written use a drug. In these situations, neither a safety study nor any effectiveness evidence is necessary. However, these examples from the published literature are necessary to comprehend the efficacy.
6. The medications from Ayurveda, Siddha, and Unani are mentioned in 3(a) as being utilized for novel indications: In these instances, the substance is as

stated in the text, but the uses are new. The demonstration of effectiveness is essential, not necessarily the safety data. Additionally, previously published literature is necessary to comprehend the efficacy.

7. There are differences in the terms relevant to safety studies for patented and non-patented medicines.
8. If any patent or proprietary medicine is listed as an ingredient in the text with a rationale for its uses, no safety study is necessary, but it is required to provide published literature as proof of its efficacy. Additionally, a pilot study is needed in accordance with the appropriate protocol for ASU drugs.

Drugs from the Ayurveda, Siddha, or Unani traditions that contain any of the components listed in Schedule E(1) of the Drugs and Cosmetics Act of 1940 have already been described in terms of their contents and intended uses. There is no additional mandate to carry out a safety study or provide any evidence of efficacy.

9.3 PROVISION OF SAFETY AND PROMOTERS OF GOOD HEALTH, SUCH AS *BALYA, POSHAK, MUQAWI*, OR *UNAVUPORUTKAL* FORMULAS

The following documentation is needed.

1. It is necessary to provide a hard copy of the textual documentation for each ingredient used in the composition as listed in any ASU authority books included in the first schedule.
2. According to the standards for evaluating Ayurvedic, Siddha, and Unani drug formulations, data from safety studies are needed if the product contains any of the components listed in Schedule E (1).
3. If the usage is mentioned in the textual indications, the safety and efficacy research is not required.

9.4 OFFERING ORAL, SKIN, BODY, OR HAIR SAFETY WITH FORMULAS REFERRED TO AS *SAUNDARYA PRASADAK, AZHAGH-SADHAN*, OR *HUSANE AFZA*

The following documents, which are comparable to those mentioned earlier, are required.

1. It is necessary to provide a hard copy of the textual documentation proving that the components employed in the formulation were included in any ASU authority's records as per the first schedule.
2. According to the standards for evaluating the formulations of Ayurvedic, Siddha, and Unani drugs, data from safety studies are needed if the product contains any of the components listed in Schedule E (1).
3. If the usage is mentioned in the textual indications, the safety and effectiveness research is not required.

9.5 PROVISION OF DRY OR WET MEDICINAL PLANT EXTRACTS THAT ARE SAFE, INCLUDING AQUEOUS OR HYDRO-ALCOHOL EXTRACTS OR THOSE BRANDED AS *AUSHADH GHANA*

1. Aqueous plant extract from any ASU medication, using the text's specified indications without changing the component. Such situations do not call for safety assessments or any kind of proof of effectiveness.
2. Aqueous plant extract of any ASU drugs without any alternation of ingredient which is mentioned in the text but new indications of use. Such cases do not require safety studies because ingredients are already established but are required to establish the effectiveness by proof or experiment.
3. Hydro-alcohol plant extract of any ASU drugs without any alternation of ingredient and indications of use which are mentioned in the text. Such cases not required safety studies or not require any establishment of the effectiveness.
4. Hydro-alcohol plant extract of any ASU drugs without any alternation of ingredient which is mentioned in the text, but indications of use are new. Such cases are required to establish the safety studies and effectiveness through proof or by experiment and published literature.
5. Other than hydro/hydro-alcohol plant extract of any ASU drugs without any alternation of ingredient and indications of use which are mentioned in the text. Such cases required to establish the safety studies (acute, sub-acute, mutagenicity, teratogenicity) and effectiveness through proof or by experiment and published literature.

9.6 TOXICITY STUDIES OF ASU DRUGS

As per ASU guidelines, toxicities are classified into two major categories which are acute and long-term toxicities or sub-acute toxicity testing.

9.6.1 Acute Toxicity

The test drug dose adjusted about 10 times of the recommended therapeutic dose and the study is performed for 25 days. The two rodent species are selected for the pre-clinical testing and rats and mice of both sexes are selected for the study. However, it is more appropriate when using two species that one of them be selected from rodents and the other from non-rodents used for toxicity testing, and rats and mice, and some regulatory agencies are accepted. Reporting the lethal doses (LD_{50}) of the test compounds in rodents and non-rodents and toxic signs is mandatory. After treatment of dose every ½, 1, 2, 4, and 24 hours up to 14 days cage-side observations including the central nervous system (CNS) stimulation or depression, salivation or any toxic signs and the severity, onset, progression, and reversibility are recorded. Other parameters such as behavioral signs include general appearance; behavioral; activity; changing of skin color, fur, eyes, or mucous membrane; body weight; food

TABLE 9.1
Selection of Doses of Test Drugs

Groupings	No. of Animals
Control Group (Vehicle treated)	10 (5M + 5F)
Therapeutic Dose (TD)	10 (5M + 5F)
Average Dose (TD × 5)	10 (5M + 5F)
Highest Dose (TD × 10)	10 (5M + 5F)

TD = therapeutic dose; M = male; F = female

intake; blood clinical parameters; hematological values; organ weights; and urine analysis; with an autopsy of viscera and tissues taken into parametric consideration.

As per the ASU guidelines, the therapeutic dose selected is according to the literature and the minimum in both sex animals is recruited for the study. The highest dose was selected as 10 times the therapeutic dose, but maximum tolerated dose selection is based on a dose at which no observed toxic effects are noted.

9.6.2 Long-Term Toxicity Test

Most long-term toxicity testing is recommended for 90–100 days for ensuring the safety aspects of long-term or prolonged exposure to traditional uses of herbal drugs. The two species—preferably one a rodent and the other a non-rodent—are selected for the pre-clinical testing. In this study systematic toxicity, carcinogenicity and teratogenicity studies are included. Further, to conclude whether a given dose would cause adverse alterations or have no impact, its dose responsive character (DRC) must be established.

A dose selection is similar of acute toxicity, and therapeutic dose is calculated accordingly relative to average dose and highest dose. A minimum of 24 animals, equally in both sexes, are assigned to four groups as described in acute study: control group or vehicle treated, therapeutic dose (TD), average dose (TD × 5), and highest dose (TD × 10). The experiment should be done in two rodent animals, e.g., rats and mice. The parameters like behavioral signs including the general appearance; behavioral activity; changing of skin color, fur, eyes, mucous membrane body weight, food intake, blood clinical parameters, hematological values, or organ weights; urine analysis; and autopsy of viscera and tissues are evaluated after immediate exposure and post-exposure experiments. Therapeutic dose should be continued for 15 days and after 15 days of exposure, 50% of the animals are sacrificed within 48 hours, which is known as immediate exposure, whereas post-experimentation is done on the 31st day following initiation of experiment, animals are sacrificed, and the parameters are evaluated.

According to use of clinical parameters and for the duration of treatment, the pre-clinical toxicity is designed as is summarized in Table 9.3.

TABLE 9.2
Experiment Procedures in Long-Term Toxicity

	Experimental Design		Experiment Duration		
Species	**Group**	**Animals No.**	**Test Compound**	**Immediate Experiment**	**Post Experiment**
Mice (Age 4–6 wk, 20–30 g weight)	Control group or vehicle treated	12 (6M + 6F)	15 days oral	50% 15th day	50% 30th day
	Therapeutic dose (TD)	12 (6M + 6F)			
	Average dose (TD × 5)	12 (6M + 6F)			
	and highest dose (TD × 10)	12 (6M + 6F)			
Wistar Rat (Age 4–6 wk, 120–150 g weight)	control group or vehicle treated	12 (6M + 6F)	15 days oral	50% 15th day	50% 30th day
	therapeutic dose (TD)	12 (6M + 6F)			
	average dose (TD × 5)	12 (6M + 6F)			
	and highest dose (TD × 10)	12 (6M + 6F)			

TABLE 9.3
Duration of Clinical Study and Requirement of Toxicity Study

Expected Duration of Clinical Use	Requirement of Toxicity Study
One week or less than clinically use in single or repeated administration.	2 weeks to 1 month
Between one week and four weeks' use in repeated administration.	4 weeks to 3 months
Between one- and six-months' use in repeated administration.	3–6 months
More than six months or greater and/or long-term repeated use in repeated administration.	9–12 months

9.6.2.1 Behavioral and General Changes

After administration of the drug, the behavioral changes are observed, including CNS stimulation, CNS depression, grip strength, and sign of convulsion the activity, changing of skin color, fur, and salivation observed to experimental animals in each day. The features of stimulations and depressions are included in Table 9.4, and each point should be mentioned. Also, visual examination, including eyes and mucous membrane body, and auditory examination should be included in the test battery, and

TABLE 9.4
The Stimulation and Depression Signs in Experimental Animals

Stimulation (CNS)	Depression (CNS)
Hyperactivity	Sedation
Piloerection	Loss of pinna
Twitching	Reflux catatonia
Rigidity	Ataxia
Irritability	Loss of muscle tone
Jumping clonic	Analgesics
Convulsions	
Tonic convulsion	
Ptosis	

the test should be done in fixed number of animals during the administration period. In CNS, stimulation and depression are observed and scaled accordingly.

Body weight: Weekly basis body weight should be recorded before the starting of drug administration and after the treatment.

Food intake: Weekly food intake in grams/day should be recorded before the start of drug administration and after the treatment.

In the event that the test substance is supplied in conjunction with the food, it is recommended to assess the consumption on a weekly basis.

9.6.2.2 Hematologic Examination

Hematological parameters and derived parameters such as the pact volume, hematocrit value, etc., should be determined before the experiment and after the experiment. Hematological toxicity is considered one of the major inclusive criteria of toxicity. Most of the toxic chemicals or herbal drugs show toxicities by hematological alternations. The blood is collected before autopsy in case of rodents, whereas non-rodent blood samples should be taken before the experiment studied and at least once during the administration period, and finally before autopsy. The hematological parameters are to be considered viz. final bone marrow cell count, complete blood count, red blood cell count, white blood cell count, hemoglobin level, and platelet count, reticulocyte count, and erythrocyte sedimentation rate (ESR) is to be considered for non-rodents only. Also, consideration is given to coagulation parameters such as prothrombin time, activated partial thromboplastin time, bleeding time, and coagulation time.

9.6.2.3 Biochemical Parameters

Major herbal drugs are metabolized and excreted through two vital organs, liver and kidneys and these two organs are to be considered the most viable in production of toxicities. Alternation of the biochemical marker of the two organs is likely to conclude the toxicity and the functions of liver and kidney should be monitored in long-term toxicity studies. Tests are performed during drug administration, and at least once during the administration period for rodents. The various parameters are tested during the toxicological studies, which include liver function test marker enzymes—blood electrolytes: sodium, potassium, phosphorus, and calcium; blood urea nitrogen, creatinine, total proteins, albumin, and globulin and other markers—glucose, cholesterol, triglycerides, HDL cholesterol (non-rodents only), and LDL cholesterol (Non-rodents only).

9.6.2.4 Urine Analysis

The metabolites and unchanged drugs are excreted through the urine, and alternation of urine compositions indicates the toxicity. The various parameters are examined, including color, appearance, specific gravity, 24-hour urinary output reaction (pH), albumin, sugar, acetone, bile pigments, urobilinogen, and occult blood.

9.6.2.5 Organ Weight

Alternation of the organ weight is also indicative of the toxicity and after experiments, weights are calculated for the major organs, including lung, brain, adrenal, heart, liver, kidney, testes, and uterus.

9.6.2.6 Gross Examination of Organs

Abnormalities of the tissue and bleeding are investigated after experimentation. Vital organs are taken to consideration, including brain cerebrum, cerebellum, midbrain, skin, liver, kidney, spinal cord, eye, heart, duodenum, middle ear, rectum, epididymis, urinary bladder, spleen, testes, mammary gland, pancreas, aorta, jejunum, ovary, mesenteric lymph node, terminal ileum, colon, uterus, esophagus, stomach, trachea, etc.

9.6.2.7 Recovery from Toxicity

Any sign of toxicity is observed, and definite periods are watched. Recovery from the toxicity signs and any toxic changes are also observed with a specific time after administration of Ayurvedic medicine.

9.7 REPRODUCTIVE AND DEVELOPMENT TOXICITY STUDIES

The ASU drugs under the section 3 (h) include crude drugs, aqueous and hydroalcoholic extracts(s), or any other solvent based extract(s) or ingredients of Schedule E (1) which are used for treatment for women of child-bearing age. The reproductive toxicity studies are divided in three categories: female fertility study (Segment I), teratogenicity study (Segment II) and perinatal study (Segment III)

9.7.1 Female Fertility Study (Segment I)

In this study, a minimum of one species of rodent like mice or rats are used of both male and female. Ideal for this study is a number of 20 males and 20 females of minimum of 40 days of age are used in one group test; the drug should be administered before minimum of 14 days before mating and up to gestation period. The route administration of the test met aerial should be similar to the clinical purposes; the dose level is selected to low, medium, and high doses; and the data are compared with the control group. The successful copulation requires a minimum two-week mating interval between male and female couples or the housing of paired animals prior to the two-week mark. After confirmation of copulation, the mortality, general signs noted, body weights, and food intake are to be noted. The effects of Ayurvedic medicine on fertilization, cycles, and autopsy are conducted for examination of genitival organs like corpora lutea, uterus, all dams, etc., as well as noting the successful pregnancies, litter sizes, numbers, and mortality of fetuses. Unsuccessfully copulating males are sacrificed and autopsied, and reproductive tissues observations are made.

9.7.2 Teratogenicity Study (Segment II)

In this study, the how the toxicity affect in organogenesis and known as teratogenicity study. Similar to the female fertility study (Segment I), in this study also a minimum of one species of either female rodent or a non-rodent is used. In this study, a minimum of 30 rats or mice and 12 rabbits should be in each group. This study also maintained a positive group with known reproductive toxicity, another negative control where no drug is given, while the study group or the route of drug administration should be similar as prescribed through clinical route. Another group is deployed of similar drugs which are known as a comparative control group. An additional group may be used for animals having a similar chemical structure or pharmacological effects of the tested drug to predict the safety of the tested drugs.

Similar observations are to be noted after confirmation of copulation. The mortality, general signs noted, body weights, and food intake are to be noted. The effects of Ayurvedic medicine in fertilization, cycles, and autopsy are conducted to two-thirds of dam in the rodent group, while all dams of the rabbits for examination of genitival organs like corpora lutea, uterus, all dams, etc.; also, note the successful pregnancies, litter sizes, implantation sites, numbers, and mortality of fetuses. Reaming of one-third of dams are allowed to deliver the litter and observed any abnormality to dam during the delivery. Further, litter body length, any abnormality in visceral or skeletal of litter, gender of litter, and gross observation organ should be made from dams. First generation litter or offspring observed any abnormality, growth sign, and reproductive performance to the abnormality to establish pregnancy. Further, autopsies are to be performed of the offspring to conclude the gross observation of the organs and tissues of treated dams, and if required second generation litters should be done.

9.7.3 Perinatal Study (Segment III)

In this segment, the toxic effects of test drugs are evaluated in perinatal and lactation phases. The similar species of Segment II or teratogenicity study are selected. The group is selected to negative control and positive control whereby the reproduction toxicity is established, and comparable structure or pharmacology of the similar test drugs are desirable, while the test drugs are taken in three doses in the consideration at high, low, and middle dose. A minimum of 20 animals is required top each group. All dams are carefully observed, including body weight, food intake, mortality, and general signs. Further dams are allowed to observe their nursing and abnormal delivery. Further, mortality, sex, length, body weight, and external changes of litter size are determined. The offspring are to be observed carefully about their growth, any abnormality of the skeleton muscle behavioral changes, and reproductive function.

9.8 LOCAL TOXICITY TESTING OF TOPICAL PREPARATIONS OF ASU DRUGS

According to section 3(h) of the Drugs and Cosmetics Act of 1940, hydroalcoholic extracts or crude aqueous extracts are any textual justification or any solvent extract of ASU medicines or ingredients of Schedule E (1) of the Drugs and Cosmetics Rules, 1945, ASU drugs are required to undergo local toxicity investigations which are to be applied topically, and the skin or vaginal mucous membrane irritation studies are prescribed.

9.8.1 Dermal Toxicity Study

Rabbit and rat are prescribed animals for sub-chronic (7 to 90 days) studies with similar to clinical dosage as applied to the dermal studies in daily basis. The dosage forms are prepared suitably, are applied and adhere to the shaved skin of a minimum of 10% of the body surface with porous gauze. Solid test material is prepared to be adhered to skin, whereas liquid and semi-solid preparations are fixed without dilution of the sample. The formulations should be at the three concentrations and the maximum concentration in actual of 10 times the clinical dosage. The signs of toxicity are noted formations like eschar formation, erythema, and edema, and further histological examinations are to be done of the damaged tissue.

9.8.2 Photoallergy or Dermal Phototoxicity

This is known as Armstrong–Harber test used in guinea pig modeling: the drugs which are photosensitive or the nature of action like treatment of leucoderma or any melatonin or pigmentation of diseases. Initially, eight animals are used with the four concentrations of test material placed to the patch and exposure of UV at the dose of 10 J/cm^2. The observation made from 15 minutes to two hours to confirm if toxicity is shown. The dose could be established from the pre-test and the concentration of active agents.

The final test is carried out with ten test animals, and additionally, five animals are used as control. The test conditions are to be similar to the pre-test conditions and repeated the experiments on days zero, two, four, seven, nine, and eleven of the tests. Final tests are to be done within 20–24 days for two hours' application, followed by allowed to exposure 10 J/cm^2 of ultraviolet light to conclude the highest dose which is a no-irritant dose. Positive control is used as musk ambrett or psoralin and comparative studies are done.

9.9 VAGINAL TOXICITY TEST

The semi-solid preparations like pessaries, suppositories, creams, or ointments are evaluated by topical application to vaginal mucosa. Rabbits or dogs are used for this study and six or ten animals are taken in a group for seven days to a maximum of 30 days. Observations are based on irritation, closure of introitus swelling, and histopathology of vaginal wall to be detrained.

9.10 RECTAL TOLERANCE TEST

This test also applied to the semi-solid's preparations like pessaries, suppositories, creams, or ointment are evaluated by applying topically to the rectal cavity. Similar to the previous study, rabbits or dogs are used for this study and six or ten animals are taken in a group for seven days to a maximum of 30 days and applied several times as equivalent to multiple of minimum of six times of human dose. The clinical signs are taken consideration like signs of pain, bleeding, sliding of backside, symptoms to the anal region/sphincter region, and microscopy or histopathology of rectal mucosa.

9.11 ALLERGENICITY/HYPERSENSITIVITY REACTIVITY

Two test models are prescribed, guinea pig maximization test (GPMT) and local lymph node assay (LLNA), and one of the tests is required.

9.11.1 Guinea Pig Maximization Test (GPMT)

In the initial experiment, the maximum non-irritant and minimum irritant doses are determined at the four dose levels by intradermal induction. A total of 32 animals, divided equally between the sexes, are divided into four groups and administered four dose levels to establish the maximum and minimum dose levels. Among the divided group, Freund's adjuvant is inoculated into two animals of each sex.

At the final experiment, a total 36 animals, divided equally between the sexes, are split into three groups, i.e., test, control, and positive control. Initial intradermal induction (day 1) is followed by topical challenge conducted after 21 days, and if no response is noted, again at 28 or 30 days. The reaction is noted, like erythema and edema scores of individual animals noted in a maximization grading scale.

9.11.2 Local Lymph Node Assay

This is an indirect method and estimated the cell proliferation of lymph nodes by 3H-thymidine or bromodeoxyuridine (BrdU) and the basis of increase in 3H-thymidine or BrdU incorporation. To test drugs, positive control is employed either sex (six animals each group) of mice on ear skin in three graded doses from low to high and a non-irritant dose, plus a vehicle control group is also allocated. It is recommended to utilize a minimum of 6 mice per group. The test material should be administered topically to the skin of the ear for three consecutive days. On the fifth day, the auricular lymph nodes should be examined for drainage and subsequently dissected for further analysis for the 3H-thymidine or bromodeoxy-uridine (BrdU) incorporation assay.

9.11.3 Other Local Toxicity Tests

Specific toxicity tests are required for the ASU drugs to be used on locally, like inhalation toxicity tests, ocular toxicity tests, etc.

9.12 SPECIAL TOXICITY STUDIES

The ASU drugs or any constituents are secluded in E (1) which are to be used for chronic use, while any drugs to be used more than six months are regulated by rules 158-B under section 3(a), 3(h), of Drugs and Cosmetics Rules, 1945. Additional tests—carcinogenicity and genotoxicity—are required for the approval process.

9.12.1 Carcinogenicity Test

This test is usually performed on two species, preferably using the rats. The test is designed in two phases. The first phase is known as a preliminary study which is used for determination of the highest dose, and this dose is used for the entire carcinogenicity test. This phase uses ten animals of each sex (age less than six weeks), while non-rodent three animals of each sex in each group are given repeated doses for 90 days. Mouse models are avoided because of their very high incidence of spontaneous tumors otherwise to be mentioned with justification for rationale of using these animals. The route of administration shall be similar to actual clinical use, and three dose levels are used. In the preliminary study, only body weight and food and water consumption are noted. Those animals which died during the experiments are allowed for the microscopy, histopathology, and gross observation of anatomy, and the determination of the possible reasons for death are to be recorded. In this stage, those animals which exhibit 10% of the weight gain, with comparison of control and no death or no toxicity was found are known as the maximum of dosing group.

Phase II or full-scale study where large animals are taken for the experiment as compared to Phase I. A minimum of 50 animals of each sex are taken and arranged for 18–24 months of study. The route of administration is to be similar to clinical use. The observation during the experiment is noted minimum weekly parameters like average of food consumption, weight changes, and any behavioral changes. The clinical biochemistry encompasses various diagnostic tests, such as glucose, lipid profile, kidney and liver functioning testing, as well as electrolyte analysis are to be determined to ensure any alternation of the physiological changes, and this test is to be performed before and after the experiments. Also, the hematologic examination—including hemoglobin content, complete blood picture, and differential count—is performed initially at the end of the experiment. A gross observation sign of any bleeding of the main organs and cavities are observed within 48 hours of the last dose after sacrificing. Further, gross histology of the main organs is studied to understand the severity or degree of any changes because of toxicity. The final conclusion is made on the basis of development of tumors, frequency of tumors, formation of tumors in the particular organ (if any), and the comparison study of the control and test group as to the degree of tumors formations.

9.12.2 Genotoxicity Tests

These are used to assess the effects of ASU drugs, whether these effects occur directly or indirectly, to produce the genotoxicity and the experiment data are required before any Phase III clinical trial. Genotoxicity studies are conducted in both *in vitro* forms and *in vivo* forms. *In vitro* tests are conducted to gene mutation in bacteria or resistance of thymidine kinase (tk) in mouse lymphoma tk (+/–) cells. The other experiment in *in vivo* testing is performed to determine chromosomal damage in rodent hematopoietic cells.

9.12.2.1 Ames' Test, Known as Reverse Mutation Assay

In vitro tests are known as Ames' Test or known as reverse mutation assay in *Salmonella.* The strains of *Salmonella* like TA98, TA100, TA102, TA1535, and TA97 or *Escherichia coli* WP2 uvrA or *Escherichia coli* WP2 uvrA (pKM1O1) are used. The positive control whereby the genotoxicity induced by sodium azide or 9-aminoacridine, 2-nitrofluorine or mitomycin C and the experiment are performed with or without the metabolic activator S9 mix. A test of ASU drugs increasing revertants numbering 2.5-fold (or more) with comparison of spontaneous revertants would be considered positive.

9.12.2.2 *In Vitro* Cytogenetic Assay

This test is similar to the previous experiments. The component of ASU drugs or ASU drugs used to determine toxicity by *in vitro* cytogenetic tests using Chinese hamster ovary (CHO) cells or on human lymphocytein cell lines are considered safe if it reduces less than 50% of cell numbers or culture confluency. The lymphocyte assay are considered safe if mitotic index reduces less than 50% when compared with the positive control used as cyclophosphamide with metabolic activation and mitomycin C. The minimum of three replicants are used and aberrations of chromosomes in metaphase are considered evaluation criteria.

9.12.2.3 *In Vivo* Micronucleus Assay

Mice are used in this assay and divided in minimum of three dose levels of ASU drugs, or any component of ASU drugs, and each group contains a minimum of five animals. Besides these, positive controls are used which are injected with mitomycin C or cyclophosphamide. The dosings are done two days in repeated doses, while before sacrificing of the animals, the last dose are given before six hours. The bone marrow is flashed out from femoral bone with fetal bovine serum and pelleted, followed by smearing them on glass slides with staining with Giemsa–May–Grünwald stain. The polychromatic erythrocytes or micronuclei formation (minimum 1,000) is taken as considered the evaluation criteria.

9.12.2.4 *In Vivo* Cytogenetic Assay

In this assay, rodent species—preferably rats—are used and divided in minimum of three dose levels of ASU drugs or any component of ASU drugs, and each group contains a minimum of five animals (either sex of animals). A positive control is used which are injected with cyclophosphamide. The dosing is done in two days. On the first day, after 22 hours of intra-peritoneal administration of colchicines, cyclophosphamide is injected, while before two hours of sacrificing of the animals, the last doses are administered intra-peritoneally. The bone marrow is flashed out from femoral bone with hypotonic saline and pelleted, followed by being resuspended in Camoy's fluid. The pellets are smeared are on glass slides with staining with Giemsa–May–Grünwald stain, and aberrations in metaphase chromosomes are observed in a minimum of 100 cells.

BIBLIOGRAPHY

1. Central Council for Research in Ayurvedic Sciences, Ministry of AYUSH, Government of India (2018). General guidelines for safety/toxicity evaluation of Ayurvedic formulation. Guideline Series-1. New Delhi.
2. Central Council for Research in Ayurvedic Sciences, Ministry of AYUSH, Government of India (2018). General guidelines for safety/toxicity evaluation of Ayurvedic formulation. Guideline Series-II. New Delhi.
3. Bhattacharya R, Sahu M, Sharma V, Shukla SS, Pandey RK (2019). Recent Advancement in In-Vivo and In-Vitro Toxicity Studies for Ayurvedic Formulation. *Indian J of Pharmaceutical Education and Research* 53(3):366–375.
4. Official Gazette vide notification No. F.28–10/45-H(I) dated the 21 December, 1945 and the last amended vide No. GSR 602(E), dated 19–7–2010.
5. Official Gazette vide notification No. K.11020/02/2010-DCC (AYUSH) dated the 21 December, 1945 and the last amended G.S.R. 663 (E) Drugs and Cosmetics (6th Amendment) Rules, 2010, Ministry of Health and Family Welfare (Department of Ayurveda, Yoga and Naturopathy, Unani, Siddha and Homoeopathy [AYUSH]) vide No. GSR 602(E), dated 10 August 2010.
6. OECD (1983). Test Guideline 415. One Generation Reproduction Toxicity Study. In: *OECD Guidelines for the Testing of Chemicals.* Organisation for Economic Cooperation and Development, Paris.
7. OECD (2001). Test Guideline 416. Two Generation Reproduction Toxicity Study. In: *OECD Guidelines for the Testing of Chemicals.* Organisation for Economic Cooperation and Development, Paris.
8. WHO (1990). *Principles for the Toxicological Assessment of Pesticide Residues in Food, Environmental Health Criteria 104.* IPCS/WHO, Geneva.
9. OECD (2001). Test Guideline 420. Acute Oral Toxicity—Fixed Dose Procedure (FDP). In: *OECD Guidelines for the Testing of Chemicals.* Organisation for Economic Cooperation and Development, Paris.
10. OECD (2001). Test Guideline 423. Acute Oral Toxicity—Acute Toxic Class Method. In: *OECD Guidelines for the Testing of Chemicals.* Organisation for Economic Cooperation and Development, Paris.
11. OECD (2006). Test Guideline 425. Acute Oral Toxicity—Up and Down Procedure (UDP). In: *OECD Guidelines for the Testing of Chemicals.* Organisation for Economic Cooperation and Development, Paris.
12. OECD (1995). Test Guideline 407. Repeated Dose 28-day Oral Toxicity Study in Rodents. In: *OECD Guidelines for the Testing of Chemicals.* Organisation for Economic Cooperation and Development, Paris.
13. OECD (1998). Test Guideline 408. Repeated Dose 90-day Oral Toxicity Study in Non-rodents. In: *OECD Guidelines for the Testing of Chemicals.* Organisation for Economic Cooperation & Development, Paris.

14. Paynter OE (1984). *Oncogenic Potential: Guidance for Analysis and Evaluation of Long-Term Rodent Studies. Evaluation Procedure #1000.1*. Office of Pesticide and Toxic Substances, EPA, Washington, DC.
15. Chandramouli R, Thirunarayanan T, Mukeshbabu K, Sriram R (2010). Designing Toxicological Evaluation of Ayurveda and Siddha Products to Cater to Global Compliance—Current Practical and Regulatory Perspectives. *Journal of Pharmaceutical Sciences and Research* 2:867–877.
16. OECD (Organisation for Economic Cooperation and Development) (2009). Carcinogenicity Studies. OECD guidance 451 adopted 7–09–2009.
17. OECD (Organisation for Economic Cooperation and Development) (2009). Combined Chronic Toxicity/Carcinogenicity Studies. OECD guidance 453 adopted 7–09–2009.
18. OECD (Organisation for Economic Cooperation and Development) (1987). Acute Dermal Toxicity. OECD guidance 402 adopted 24–02–1987.
19. OECD (Organisation for Economic Cooperation and Development) (1984). Genetic Toxicology: Rodent Dominant Lethal Test. OECD guidance 478 adopted 4–04–1984.
20. Vahalia MK, Thakur KS, Nadkarni S, Sangle VD (2011). Chronic Toxicity Study for Tamra Bhasma (A Generic Ayurvedic Mineral Formulation) in Laboratory Animals. *Recent Research in Science and Technology* 3(11):76–79.
21. Bhattacharya R, Sahu M, Sharma V, Shukla SS, Pandey RK (2019). Recent Advancement in In-Vivo and In-Vitro Toxicity Studies for Ayurvedic Formulation. *Indian Journal of Pharmaceutical Education and Research* 53:366–375.

10 The Embarking of Food Safety and Standards Authority of India (FSSAI) and Implantation of the Food Safety and Standards Act

10.1 INTRODUCTION

Under The Ministry of Health and Family Welfare's Food Safety and Standards Authority of India FSSAI is an independent statutory organization that oversees food safety and standards in India.

The FSSA, 2006, governs the FSSAI and its function. The aforementioned statute has been strengthened and governs the production, distribution, sale, and import of foods that may be considered wholesome and safe for human consumption. The FSSAI combined many legal frameworks—such as those governing food production, quality assurance, standards, and laws prohibiting food adulteration—into a single, comprehensive law. The goal of the FSSAI is to establish regulations via the food industry as a whole to ensure that safe, healthy food is provided. In addition, the FSSAI established guidelines for the requirements for food quality, certification, and accreditation by informing the labs and facilities. Additionally, FSSAI controls food contamination, residues, and developing risks. FSSA 2006 8 August 2019, the FSSAI began enforcing new modifications to the Food Safety and Standard Act (FSSA), 2006 regarding food safety, including any pollutants, toxins, and residues.

10.2 INTRODUCTION TO THE FOOD SAFETY AND STANDARDS ACT

The Indian government is concerned with protecting consumer rights and putting an end to the use of unsafe, poisonous, and unclean foods. This is the rationale behind the implementation and enforcement of laws. Particularly, packaged goods were subject to government regulation in India. At first, several statutes and orders were put into effect, and numerous independent agencies also established the benchmark standards for food quality. India adopted the following laws and acts.

DOI: 10.1201/9781003398431-10

The Prevention of Food Adulteration Act was passed in 1954.
The 1955 Fruit Products Order.
The 1973 Meat Food Products Order.
The 1947 Vegetable Oil Products (Control) Order.
The Edible Oils Packaging (Regulation) Order, which was passed in 1998.
The 1967 Solvent Extracted Oil, Deoiled Meal, and Edible Flour (Control) Order.
The 1992 Order for Milk and Milk Products.
The 1955 Essential Commodities Act (in relation to food).

The Prevention of Food Adulteration Act of 1954, which went into effect on 15th June 1955, is one of the most notable laws. It was designed to ensure that customers knew whether foods were healthy or unhealthful. Additionally, this regulation serves as a safeguard to end food producers' mall practices and require them to engage in fair trade commerce. By two amendments made in 1964 and 1976, the law's scope of punishment was put into effect. The Fruit Products Order was also issued by the Indian government in 1946 and revised in 1955. A legal requirement was established to guarantee the minimum quality of goods and vegetables. Additionally, the two standards are the AGMARK standard and the Indian Standards Institute (ISI) standard implemented. Oversight of goods distributed throughout India is the ISI standard, which was developed by numerous committees, governmental organizations, and a quality control laboratory. The Directorate of Marketing and Inspection of the government of India developed the AGMARK standard, which is primarily used for agricultural products to increase consumer confidence in their safety and purity.

An enormous amount of foreign goods is being harbored by the expanding food market and the open market philosophy. The FSSA, 2006 is the statutory and regulatory framework for pre-packaged goods. The government of India recognized the need for a standardized set of guidelines and created the regulation to bring all food-related laws under one roof. The FSSA acts replace one legislation and eight previously enacted laws. The (FSSAI) is a national regulating body for the food industry that was founded to conform with international standards.

The Act is required to govern all trading and commerce involving food or substances connected to food, including production, processing, distribution, storage, sale, import, or export, etc., in order to maintain food safety standards. On 5th August 2011, the Indian government issued a notification and began implementing it. The FSSA, 2006 implemented stringent regulations prohibiting non-scientific, unsubstantiated claims from being included on labels.

The FSSA, 2006 does not possess any of the statutory or regulatory obligations, but certain provisions do regulate advertising.

10.3 THE FOOD SAFETY AND STANDARDS ACT'S IMPORTANT SECTIONS ON FOOD LAWS

1. Regulation 2.12: Clearly explains the definition of "proprietary food" and the exclusion of such foods from this Act under this section.

2. Section 1 (8): According to this act's packaging and labeling regulations, "pre-packaged" or "pre-packed food" is taken into consideration. All packaged foods are governed by this clause.
3. Section 22: All foods with value-added properties, such as nutraceuticals, functional foods, novel foods, and health supplements, are categorized here. All of the foods in this category must have scientific support for their claims. If these products make claims to offer nutritional or therapeutic benefits, they must be supported by credible scientific data.
4. The Indian Council of Medical Research's recommended dietary allowances for Indians must be met by the products' vitamins and minerals, which should be free of steroids and psychotropic substances. FSSAI, however, asserted section 22 by the Honorable Supreme Court's in 2015 directives, and a change is being considered or submitted.
5. "Advertisement" was defined in the FSSA act's Section 3 (1) (b), and this section applied equally to "notice".
6. The claimant is shown in the format of employing light, print, electronic media, sound, smoke, gas, internet or websites, or any other methods of notice, label, wrapper, circular, invoice, or other papers, in accordance with this section.
7. According to Section 24 of the FSSA, advertisements that are false, deceptive, or violate other laws should not be published.
   a. The guidelines and requirements for advertising are described in this section.
8. This clause also instructs individuals not to engage in unfair trade practices, such as boosting the supply, usage, and consumption of any food particles.
9. This part also addressed the avoidance of misleading practices, such as the use of any illogical or unsupported claims made in writing, orally, or visually that serve to advertise the sale of food items.
10. The claim that a food product is distinctive should be supported by scientific evidence, such as its standard, quality, grade composition, and usefulness. Additionally, any assertions are taken into account as a promise to the public and should not be considered a sufficient or scientific basis. Bark is examined under a microscope to check for the presence of starch, tannins, or lignified components. The thin section of the roots and rhizomes is examined under a microscope to check for the presence of starch, inulin, lignified components, or fixed oil. The claimant should be able to support his or her position with the appropriate documents, academic research, and scientific evidence. Any infraction or untrue statement may constitute a criminal offense. Under the Packaging and Labeling Regulations,
11. The Food Safety and Standards (Packaging and Labeling) Regulations, 2011, were implemented by the FSSAI to ensure uniformity and readable printed basic label material for pre-packaged, proprietary, and other particular items.
12. Although the label may vary depending on the type of food, the core information is always there.

13. This rule mandates the bilingual language declaration in both English and Hindi written in Devnagri script.
14. Although the label may vary depending on the type of food, the core information is always there. This rule mandates a bilingual language declaration in both English and Hindi written in Devnagri script.
15. Nothing in the written material can be misleading, deceptive, or give the wrong impression, and all assertions are supported by scientific evidence.
16. Laminated printing is allowed so that pre-packaged food labels can be kept apart from the container. Label printing should be legible to consumers, conspicuous, and permanent.
17. The container's wrapper should have clearly visible information.
18. Additional significant data should be mentioned: the food's name, a list of its ingredients, and nutritional data, a statement and marking indicating if something is vegetarian or not, net weight quantity, lot/code/batch identification, manufacturing or packing date, complete manufacturer's address, "best before" or "use by" dates, the nation of origin if food is imported, and useful instructions.

10.4 ENFORCEMENT OF THE FOOD SAFETY AND STANDARDS ACT

Since the FSSA is mandated by the national government, everyone who operates a food business in the nation is required to follow it. It is also true that the state government, which is the FSSA's implementing authority, is responsible for its proper execution. The following legislations are also reported to be implemented along with the FSSA.

1. Permits from the municipal corporation for trade and health.
2. Environment clearance.
3. A certificate in fire safety and prevention.
4. Registration under the shops and establishments act of the relevant state.
5. Registration under the police act of the relevant city/state.
6. Validation certificates under the Standards of Weights and Measures Act, 1976, provided by the Department of Legal Metrology of the relevant areas for each of the outlets.
7. Registration under the restaurant license and liquor license.
8. Other licenses, such as those for operating musical instruments, are also necessary for operating a restaurant. Additionally, necessary insurance requirements for the food industry include coverage for public policies set forth by the government, such as product liability, building, fire policy, and assets.
9. Any food business must adhere to labor law. According to the Employees' Provident Funds and Miscellaneous Provisions Act of 1952, licenses and registration are necessary for each firm employing more than 20 people.

10. The third schedule of the Central Excise Act of 1944 also applies to any changes in retail sale price, repackaging, or labeling for this product, which is regarded as a category of manufacture.
11. Additional related rules and regulations include the Income Tax Act of 1861, Goods and Services Tax (GST) act 1999, Customs Act of 1962, service tax, and various labor laws.

10.5 FOOD IMPORT

1. Under the 2011 Food Safety and Standards (Packaging and Labeling) Regulation, food importation is likewise covered by the FSSAI.
2. It is necessary to specify where the imported foods are from.
3. The importing nation is regarded as the food's place of origin in the event that the raw material is imported into the importing nation and modified appropriately.
4. Food additives in imported goods must adhere to FSSA, 2006.
5. Many chocolate makers import chocolate, like Cadbury, Lindt, Ferrero, etc., but the quality and additives must comply with legal requirements.
6. All of the ingredients in alcoholic beverages, including beer, must be listed on labels in descending order by percentage of weight and volume.
7. Equally crucial is that the label should be an "inseparable label" rather than a sticker.
8. The following requirements for imported food must be met according to FSAA, 2006.
9. All label material must be written in legible English.
10. Foods should be labeled as "vegetarian" or "non-vegetarian" according to their classification.
11. The full address of the company where the imported food was made should be provided.
12. Mention the contents' net weight, amount, or volume.
13. Mention the lot number, code number, batch number, or FSSAI license number.
14. The day and month that the product was made.
15. The package's expiration date and the word categories with the "best before" validation should be stated.
16. It is important to include the nutritional information or nutritional intake per 100 g or 100 ml of food.
17. The label needs to include the address of the manufacturer.
18. The license number and official FSSAI logo should be displayed on the label.

10.6. FSSAI, AND INTERNATIONAL REGULATORY AGENCY

1. After the era of free trade and marketing, it became difficult to maintain food quality in accordance with international standards like Codex Alimentarius, and the FSSAI was established as a result of enforcement of codex alinement's in India.

2. The Codex alimentarius is the result of collaboration between the WHO, the Food and Agriculture Organization, and other food- and health-related UN agencies.
3. The international food standards, recommendations, and codes of conduct established by Codex Alimentarius help ensure that the food that people consume is both safe and of high quality.
4. Since the FSAA, 2006, are based on Codex Alimentarius criteria, they uphold global standards.
5. The Codex Alimentarius serves as the foundation for a number of other international standards created by international organizations as the European Food Safety Authority, Food Standards Australia New Zealand, and the FSSAI, which is authority of implementing the food safety laws in India.
6. The World Health Organization and the Food and Agriculture Organization, two premier United Nations health and food bodies, worked together to create the Codex.
7. The safety and quality of the food that is delivered to consumers is aided by the international Codex Alimentarius food standards, recommendations, and codes of conduct.
8. Since the FSSAI regulations are based on Codex Alimentarius criteria, they uphold global standards.
9. Other worldwide organizations, such as the European Food Safety Authority, Food Standards Australia, New Zealand, the FSSA, which is consisting of 12 chapters, have developed the following additional international standards:
    a. Definitions that are preliminary.
    b. India's Food Safety and Standards Authority.
    c. Basic rules for food safety.
    d. General food prohibition provisions.
    e. Import-related provisions.
    f. Specific obligations with regard to food safety.
    g. Implementation of the law.
    h. Labs, sampling, and public analysts for food analysis.
    i. Penalties and offenses.
    j. The tribunal for adjudication and food safety appeals.
    k. Audits, reports, and financial matters.
    l. Miscellaneous.
10. This Act also includes the following nine expert panels, including the Panel for Functional Foods and Nutraceuticals.
    a. Panel for Sampling and Analysis Methods.
    b. Food Additives Panel.
    c. Panel for Food Chain Contaminants.
    d. Biological Hazards Panel.
    e. Pesticide and Antibiotic Residues panel.
    f. Labeling and Claims/Advertisements Panel.
    g. Panel for Foods and Genetically Modified Organisms.
    h. Fish and Fisheries Panel.
    i. Products Panel.

10.7 FOOD SAFETY AND STANDARD (HEALTH SUPPLEMENTS, NUTRACEUTICALS, FOOD FOR SPECIAL MEDICAL PURPOSES, FUNCTIONAL FOOD) REGULATION 2016 ENFORCED IN INDIA AND SALIENT FEATURES

10.7.1 Chapter I: Preliminary: definitions

1. Foods containing non-viable food components that altered the gut flora to create positive effects are regarded as containing prebiotics.
2. "Foods containing Probiotic Ingredients" are defined as live microorganisms that are consumed with food, either as a single active component or a combination of active competent, and that has or have a positive impact on human health. Furthermore, after passing through the digestive system, this probiotic bacterium can persist in sufficient numbers, attach, and grow in the gut, producing the physiological effect.
3. Foods for Special Dietary Uses (FSDU) (apart from newborns and those to be consumed on a doctor's advice)
4. Two different food categories benefit are taken into account here.
5. The first benefit of customized processed foods is the improvement of physiological or particular health issues, such as diabetes, high blood pressure, underweight, obesity, etc. These foods are not consumed often or on a daily basis. The second benefit is foods are frequently consuming with meals that are nutrient-rich and suited for general health improvement. This kind of food is not allowed to be used with claims that it will improve disease symptoms or be directed at a particular consumer group. This kind of food is not covered by the regulations of Foods for Special Medical Purposes (FSMP).
6. The term "food or health supplements" refers to items that are consumed on a regular basis and are considered to be normal foods. These items typically contain vitamins, minerals, mineral complexes, amino acids, proteins, and enzymes.
7. Other food components have nutritional worth or have been proven to improve physiological circumstances, whether they are of plant or animal origin. The 1940 Drugs and Cosmetics Act and its regulations do not apply to these foods. Foods that fall under the FSMP category are primarily divided into two types. First, customized foods comprise specific nutrients that were employed for a special person's adsorption and excretion, as well as impaired metabolism. In this situation, modifying food would not be able to manage the typical diet. Second, foods used to manage weight by substituting for a typical diet are another option.
8. The term "nutraceuticals" refers to a specific type of active ingredient that has been isolated from natural foods and has been scientifically shown to have the ability to prevent or control disease. These foods are typically administered in unit quantities as powders or granules such as a tablet, capsule, liquid, or gel.
9. "Novel foods" are defined as foods that have nutritional benefits, such as those related to metabolism and excretion, but whose historical use as

foods or sources are undocumented. This kind of food is typically prepared employing cutting-edge new engineering techniques, technologies, or procedures.

10. "Specialty foods" are foods that contain substances from Indian traditional health systems like Ayurveda, Unani, Siddha, and others. This is done to prove that the foods are safe and in compliance with Schedule I of the Drugs and Cosmetic Act of 1940 (as well as other Standard Manuals).

10.7.2 Chapter II (Generic Foods)

1. There should be a framework that makes it clear which foods are for specific nutritional purposes, which are simply foods for a diet, and which are just general foods.
2. Additionally, the dietary composition of meals consumed for specific purposes must differ from that of generic foods.
3. The production procedure is also distinct from a typical diet.
4. Special nutritional needs foods may not be produced, packaged, sold, or marketed unless they adhere to the rules outlined in this category.
5. The specific nutritional food element that has been scientifically verified and is utilized logically in accordance with the underlying nutritional and medical principles.
6. This kind of food cannot contain any extra hormones, steroids, or psychoactive compounds.
7. The target consumer group and the disease conditions where this type of food can be helpful should be specified on the labels of this type of food, which are significant.
8. Information, content, and precautions should be included on labels, leaflets, or any other type of advertisement.
9. In this type of category, only foods that are naturally derived are to be taken into account; no items that are labeled as "food supplements", "special diets", "special dietary", or any other phrase with a similar meaning are to be specified.
10. Moreover, the Food Authority informs the Food Business Operator about the production, marketing, and safety assessment of such specialized food types.

10.7.3 Chapter III (Food or health supplements)

This chapter is important with food or health supplements.

1. The foods are used as supplement with foods and marketing in a single unit mainly as capsules, tablets, pills, and other similar forms of liquids and powders, etc. In this type foods not considered the other categories foods which having specific standard with a specific regulation.

2. This chapter describes the essential composition of foods and how they are accordingly categorized as mentioned. Any foods can be modified and added more nutrients with admixing of one or more nutrients.
   a. Schedule I: Vitamins and minerals.
   b. Schedule II: Any forms of vitamins and minerals.
   c. Schedule III: Amino acids.
   d. Schedule V: Plants or botanicals.
   e. Schedule VII: Mineral or metal.
   f. Schedule IX: enzymes as specified.
3. All nutrient content should be within the limit of the maximum allowed intake as fixed by the Indian Council of Medical Research (ICMR) or maximum dosage. The limit of intake is also considered as per the other food standards and Codex or any monograph of the World Health Organization.
4. No nutrient is permissible without establishing the safety and health benefits, and Food Authority approval.
5. The specificity and purity of the nutrients including vitamins and minerals which notified by the Food Authority from time to time if the standards are not established and the FSSAI also accepted the purity of any official monograph by international bodies such as Codex Alimentarius.
6. The Food Authority also enlists and time to time approves the specific nutrients with scientific evaluation.
7. Labeling and packaging of food supplements should comply with Food Safety and Standards (Packaging and Labeling) Regulations, 2011.
8. Without the specific requirement of the label as per the laid norms of the regulation, nobody can pack, sell, distribute, or offer or import the food supplement.
9. In the advertisement, presentation, or labeling, such word should not bear that preventing, treating, or curing a human disease or concurrence of such claim. The statutory statement should be included "This product is not intended to diagnose, treat, cure, or prevent any disease(s)".
10. The label of the food supplement should contain the words/statement as follows.
11. Legible and boldly should be written such as "FOOD or HEALTH SUPPLEMENT" and "NOT FOR MEDICINAL USE".
12. The label should be mentioned the common name or a description to identify the nature and categories of foods or health supplement sufficient to indicate the true nature of the food supplement, including the common names.
13. The content and number of nutritional substances in numerical form in descending order, as well as the physiological effects, should be mentioned on the label.
14. Recommended daily allowances as prescribed by the Indian Council of Medical Research should be included and the label should include a statutory warning "Not to exceed the recommended daily dose".
15. Also, the label should be mention "not substituted for a varied diet".
16. The warning(s), contraindication(s), or precaution(s) should be mentioned on the label.

17. The other cautionary statement "store out of the reach of children" should be mentioned on the label.
18. All additives should be within the list of schedules as mentioned in VIII (a) and VIII (e) of these regulations
19. All contaminants, toxins, and residues should be regulated as per the Food Safety and Standards (Contaminants, Toxins, and Residues) Regulation, 2011

10.7.4 Chapter IV (Nutraceuticals)

1. Definition and scope: In this chapter, the nutraceuticals and character are described. Nutraceuticals, emphasizing that they must have a defined essential composition and can be derived from various sources, including plants, microbes, or animals. Additionally, you mention that the historical use of the entire source is sometimes considered.
   Nutraceuticals are defined as substances that provide health benefits and have a role in preventing or treating diseases. They are often positioned at the intersection of nutrition and pharmaceuticals. Nutraceuticals are required to have a defined essential composition, implying that the specific components contributing to their health benefits are identified and considered essential to their function.
   Nutraceuticals can originate from various sources, including both food and non-food sources. This includes plants, microbes (microorganisms), and animals. The historical use of the entire source is mentioned, suggesting that the traditional or historical use of the complete plant, microbe, or animal may be considered in the context of nutraceuticals. This historical aspect could be relevant for establishing safety and efficacy. Nutraceuticals are sometimes extracted as amino acids and their derivatives. This extraction process may involve fermentation foods, systems, or full-grown bacteria. Amino acids are building blocks of proteins and can have specific health benefits.
2. Average Daily Intake (ADI) should not exceed more than as specified in the Indian Council of Medical Research (ICMR). If the ADI is not mentioned in ICMR, the Food Authority should follow the international food standards Codex Alimentarius or the World Health Organization.
3. The nutraceuticals are not used in India and the safety profile is not established in India, but this type of nutraceuticals have previous history of use in other countries, and this case FSSAI can permit after taking approval. This legal requirement suggests that nutraceuticals must have a demonstrated history of safe and established use before they can be allowed for consumption and sale within the Indian market. The specific timeframes are 10 years within India and 30 years in foreign countries. This requirement is likely in place to ensure the safety and effectiveness of nutraceuticals in the Indian context.
4. The nutraceuticals lists are amended from time to time after undertaking proper scientific evaluation. The concerns of the purity and quality controls are regulated by the FSSAI as per the schedules V, VI, and IX. Also,

the official monographs/pharmacopoeias like Indian Pharmacopoeia (IP), British Pharmacopoeia (BP), and United States Pharmacopoeia (USP), and international bodies such as Codex Alimentarius are also accepted in non-availability in the schedules.

5. The labeling of all nutraceuticals should adhere to the Food Safety and Standards (Packaging and Labeling) Regulations, 2011
6. No one is allowed to produce, market, distribute, or import a package or container that contains a nutraceutical if they are not in compliance with the law.
7. The label should be truthful, avoid disease-related claims, and be backed by scientific research to provide consumers with accurate information about the product's effects and benefits.
8. The common phrase "This product is not intended to diagnose, treat, cure, or prevent any illness" should be present.
9. The word "nutraceutical" should be written in bold and be easily readable on the label.
10. The nutraceutical's common name should be mentioned.
11. It is important to mention the active ingredient in a nutraceutical that has a nutritional or physiological benefit. According to the Indian Council of Medical Research, the average daily intake should be indicated along with the phrase "not to exceed the stated recommended daily dose".
12. "Recommended usage" must be put on the label, and it is also possible to write about the negative effects of excessive use.
13. The phrase "NOT FOR MEDICINAL USE" must be included on the label. It should be noted that excessive use can be harmful.
14. A warning statement should be included regarding any potential interactions between foods, medications, or both.
15. Legible printed warnings and guidelines such as "Store in a cold area", "Keep away from children", etc.
16. The font size for all cautionary statements should be at least 3 mm (8.5 points).
17. All additives should be listed in accordance with Schedule VIII(a) and VIII(e) of the regulations.
18. Food Safety and Standards (Contaminants, Toxins, and Residues) Regulation, 2011, as amended from time to time, should be followed with regard to any contaminants, toxins, or residues.

10.7.5 Benefits of Food Claims

1. The claims made about nutraceuticals are crucial, and it is clear that they should not fall under the Drugs and Cosmetics Act of 1940. Claims are taken into account for the two groups: (i) a nutritional claim that is subject to nutritional supplement regulations and (ii) health claims on the possible improvement of the health beneficiary but not taken into account in the treatment, diagnosis, or prevention of disease.
2. The nutraceutical's constituents and boosting health are said to have a close association. A health claim is focused on either: (i) nutritional elements or (ii) the improvement of health.

3. The control of incurable diseases and the regulation of health claims. The information suggests that claims related to advertising and magical cures for incurable diseases are not allowed, and health claims are only accepted with substantial scientific support.
   a. Control of Incurable Diseases: Claims related to the control of incurable diseases fall under the category of advertising and magical cures and are not allowed. This reflects a regulatory stance on potentially misleading or unverified claims.
   b. Frequently Asked Questions: It's not clear what specific questions are being referred to, as the statement starts with "The following questions are frequently asked." If there are specific questions related to health claims or the control of diseases, more context would be needed for a detailed response.
   c. Improving Physiological Function: The statement acknowledges that claims related to improving physiological function may be considered, but it's essential that such claims are supported by substantial scientific evidence.
   d. Claims Relating to Disease Improvement but Not Cure: The distinction is made between claims related to improving physiological function and those relating to disease improvement. The latter might be permissible as long as they are not claiming a cure and are supported by scientific evidence.
   e. Claims Pertaining to Maintaining Health and Improved Immunity: Claims related to maintaining health and indicating improved immunity are mentioned. However, the statement specifies that these claims, excluding vaccines, are subject to the requirement of substantial scientific support.
4. However, the claim should not be that it delays aging claims on is acceptance.
5. Improves circumstances of healthy aging.
6. It is believed that the idea that aging can be inhibited is erroneous because aging is a natural process and a fact.
7. Claims relating to dietary issues that claim improved physiological function or a decrease in disease.
8. The FSSAI was also required to certify the claim that was unrelated to the Drugs and Cosmetics Act's purview.
9. The FSSAI must be able to assess and approve the health claims based on a sufficient degree of documentation and reliable evidence.
   a. The claim must fall within the parameters of improved function and decreased illness risk.
   b. The validity of the nutritional claims should be used to substantiate the assertion.
   c. The assertion should fall under the category of a scientific claim and be supported by both scientific and experimental facts.
   d. Nowadays, unlike with medicines, post-marketing surveillance also monitors post-market, consumer, or cohort studies in addition to the scientific literature. Statistics on eating habits, epidemiological data from India, and health advantages are all included in the comprehensive sample group.

e. The studies are also taken into account for the definition of congruent, concurrent, and consensus validity studies.
f. If consensus and congruent research are provided to establish the efficacy and safety data for humans, the claim that they favor the reduction of disease and health promoters are taken into consideration. Additionally, the relevant published and historical scientific literature must be taken into account.
g. Epidemiology studies are suggested in addition to clinical trial date for determining efficacy and safety; also taken into account are plausible statements about the better performance of the particular organ.
h. The phrase "Prevents diabetic retinopathy in the diabetic circumstances" is an example of an ambiguous claim that is not permitted.
i. Ambiguous disease-curing claims should not be reflected in the product name, such as "AIDS cure", "cancer cure", etc. This law forbids the use of any images, vignettes, symbols, or other forms of advertisement that suggest a disease's ambiguity, such as lipid profiles, electrocardiograms, cardiac arrhythmias, etc.
j. The precise category of the user group should be mentioned with the age group, gender, or vulnerable populations in the structure-function assertions.
k. The FSSAI examines new scientific knowledge on a periodic basis in response to stakeholder requests and periodically updates the claims, safety regulations, or procedure tenets and implements them in accordance with the advice of an expert committee.
l. The FSSAI regulates the health claims made about nutraceuticals, and it requires that all pertinent information—including labels and leaflets—be submitted to them.
m. Human studies are required to back up food benefit claims. Human studies are necessary for the product supporting the following claim, which is based on assertions made by humans:

10. Claims related to risk reduction or benefit for disease; the human intervention studies are required by the claimant.
11. Claims related to nutraceutical.
12. Product compatibility studies are required which are linked to health claims like composition of polyunsaturated fatty acids for a healthy heart, etc.
13. The human intervention study is mandatory when claims include "Shown", like "(Name of the product) is shown to be helpful in keeping your heart healthy or heart-healthy" or "(Name of the product) is shown to be helpful in maintaining liver functions".
14. More human intervention studies or epidemiological data on Indian populations with concurrent validity are mandatory when claims use the word "Proven", such as "(Name of the product) is proven in the management of weight".
15. The Food Authority is notified of the foods which have health claims.
16. The pre-approval requirement is mandatory where the new molecules are going to introduced and the health claim without sufficient data or adequate scientific input.

17. The review procedure of the claim: The product should comply with the comprehensive information safety and claims which data generated by the manufacturing or marketing organization. All relevant data should be evaluated by an independent scientist or expert having related qualifications and experience. The reviewer's credibility including qualifications and experience are treated as an essential part of the document.
18. Disposal of complaint: The query related to technical matter or any complaint within the purview of the Food Authority. The authority has to review the case by appointing a committee and the recommendation of the expert group, the FSSAI, may notify to stop such claim or be directed to manufacturing firms to alter the claims.

10.7.6 Chapter V (Foods for Special Dietary Uses) (excluding infants, and those to be taken under medical advice)

1. A unit pack or the unit forms including granules, tablets, pills, capsules, sachets of powder and/or other similar forms can be used in small quantities and have special dietary uses including any disease conditions other than food supplements are categorized in this rule. This type of food can have a nutritional or physiological effect in small quantities and all these types of foods are followed the described schedules, a specific standard of the described FSSA, 2006. Unlike the other chapter, this chapter has two important components: (i) essential composition; and (ii) claims in the label.
2. Essential composition and Schedules of the compositions.
   a. Schedule I: Describes the vitamins and minerals.
   b. Schedule II: Describes the forms of the vitamins and minerals.
   c. Schedule III: Describes amino acids.
   d. Schedule V: Describes plants and botanicals.
   e. Schedule VI: Describes the animal-originated nutrients.
   f. Schedule VII: Described the mineral and metal sources.
   g. Schedules IX: Described the enzymes list.
3. The nutraceuticals are not used in India and the safety profile is not established in India, but this type of nutraceuticals have a pre-history of use in other countries, and FSSAI can in this case permit them after granting approval. FSSAI may revise the nutrients list from time to time with the validation of the scientific evaluation.
4. The content of nutrients should be within the prescribed limit and not exceed the safety level.
5. Regulation of weight management food: Special foods for use mostly in weight management and weight control purposes should be followed as follows.
6. For the formula for this type of food the total energy value should not less than 800 kcal (3,350 kJ) and not more than 1,200 kcal (5,020 kJ) of the total consumption/per day, and the food can be divided into two or three equal doses for the consumption during the whole day. The diet with the replacement of this type of food is permissible and the formula of this type of

foods should be minimum of 200 kcal (835 kJ) and not more than 400 kcal (1,670 kJ) and the total energy intake in a day shall not exceed 1,200 kcal (5,020 kJ).

7. Protein: The maximum protein consumption is not more than 125 g/per day and the limit of the total energy of the value of the food is 25% and maximum is up to 50%.
8. Protein-digestibility-corrected amino acid score (PDCAAS) maximum limit is 1.0 (the reference protein).
9. The diet used for weight management where the lower protein content is increased and the PDCAAS score is limited to 0.8.
10. The food may be composed of essential amino acids, and all are L-forms and except DL-methionine.
11. Dietary fat and linoleic acid as sources of energy, along with specifications on the maximum and minimum limits of energy intake from fats in the context of the total energy content of food. Dietary fats, including fatty acids like linoleic acid, serve as important sources of energy in the diet. Fats are macronutrients that provide a concentrated form of energy when metabolized. Maximum limit for the energy intake from fats in the total energy content of food. According to the information provided, this maximum limit is set at 30% of the total energy. Minimum energy requirement from fats, and this should not be less than 3% of the integral content of the food.
12. The replacement of 100% of the foods is required, supplemented with adequate vitamins and minerals.
13. The FSSAI may from time to time established the quality criteria for the nutrients used, such as vitamins and minerals. If the criteria are not specified, the generally recognized purity criteria from international organizations such as *Codex Alimentarius* may be referred to and adopted.
14. The labeling of the category of the Foods for Special Dietary Uses is strictly regulated under the Food Safety and Standards (Packaging and Labeling) Regulations, 2011 and should specify the details of nutritional contents and the claims.
15. Regulations, particularly related to labeling and various business activities. Any violation of regulations related to labeling in the context of business activities, including repackaging, manufacturing, selling, marketing, distributing, or importing, is prohibited.
16. The statement or content should not mislead the consumer regarding the use of foods that prevent, cure, or diagnose.
17. The label shouldn't include any assertions or suggesting for the prevention, cure, or treatment of any particular disease, its diagnosis, or the disruption of the body's natural physiological functioning, whether temporarily or permanently. The claim of such medical intervention requires the pre-approval of the FSSI with scientific validation. Also, as per the guideline, this type of food contains the disclaimer that "This product is not intended to diagnose, treat, cure, or prevent any disease". The following Declaration is required for the "Food for Special Dietary Uses (FSDU)"

a. The Declaration like "Foods for (name of physiological conditions/disease conditions)". Also mentioned is that the foods are formulated and used for physiological conditions.
b. Special foods prepared for the special purpose like the statement are required to mention like "Use for weight management" but the contra-indication should be mentioned like not use for the pregnant, lactating, and nursing women.
c. The target consumer group and justification for the use of the product should be mentioned.
d. If the product is designed for a specific age group, this should also be mentioned.
e. The rational use of the nutrition used in the foods is to be mentioned. The quantity used, deleted specific nutrition, or adding nutrition is to be mentioned.
f. Caution is to be boldly mentioned relative to the side effects of side effects of excess consumption.
g. Caution should be mentioned if there are any known side effects or food and/or drug interactions. The safe level of the quantity of the nutrition is required to be mentioned.
h. Osmolality or osmolarity or on acid-base balance should be mentioned.
i. The statement "Store out of reach of children" should be mentioned.
j. Additives
k. The additives for special dietary foods should be according to Schedule VIII (b) and VIII (e) exceptionally, other than infants. Like other foods, this type of food should be free of contaminants, toxins, and residues, according to the Food Safety and Standards Regulations, 2011.

10.7.7 Chapter VI (Foods for Special Medical Purposes [FSMP])

1. This type of food is used under medical investigation with the aim to provide adequate nutrition or deficiency to patients with a special provision of the formulation. This type of foods are used for special medical purposes for using patients under treatment or disease or disorder with nutritional deficient patients. Most of these type foods are nutritionally rich and bears statutory conditions like "Not for infants" or "Use only as prescribed".
2. Categories: These types of foods classified into three types depending upon the formulation and intended to use. This type of food can be used with partial replacement of indenting daily foods.
3. Standard nutrient formulation and use according to the instructions for the requirements of nourishment for the persons. Most of this food is standard nutritionally complete foods.
4. Standard complete foods with specific nutrition used in disease, medical conditions, or disorder conditions. Also, this type of food can be used only by manufacturer's instructions without medical interference.

5. These types of foods are special category foods or nutritionally incomplete foods, whether in a normal composition or a nutrient-adapted formulation tailored to address a particular disease, ailment, or medical condition, are not appropriate for exclusive utilization as the primary source of sustenance.
   a. Amino acids as specified in Schedule III. Vitamins as per level mentioned in Schedule IV. Plants or botanicals as specified in Schedule V.
   b. Animal origin originated substances as per list in Schedule VI.
   c. Minerals as mentioned in Schedule VII.
   d. Enzymes as per Schedule IX.
6. Manufacturing of special medicinal foods used listed one or more nutrients as per regulations without prejudice to modifications. Adding additional nutritional agents should be enlisted by the FSSAI for approval after scientific evaluation of safety and physiological benefit. From time to time, the FSSAI enlists the new nutrition guidelines.
7. Regulation of Foods for Special Medical Purposes shall comply with schedule IV and also follow the terms and conditions.
8. Foods should be prepared according to the regulations and use a minimum of 450–800 kcal as the only source of energy. The minimum protein content in daily intake foods should be 50 g protein with PDCAAS (protein-digestibility-corrected amino acid score) of 1. Except for DL-methionine, all L-forms of amino acids may be energy, and essential amino acids L-forms of amino acids shall be used, except that DL-methionine may be used for improving protein quality of foods.
9. A minimum of 3 g of linoleic acid is permitted as fat content in low energy foods.
   a. A minimum of 0.5 g α-linolenic acid is recommended as fat in daily intake and the ratio of α-linolenic should be from 5–15 maintained with linoleic acid.
      i. A minimum of 50 g of carbohydrates should be present in the daily intake of food.
   b. More than nutrition can be added for special medical purposes, and scientifically one established data and technologically correct methods should be established before adding.
   c. The use and necessity of the specific group should be mentioned with cautionary labeling. Permission is required from the FSSAI.
   d. The quality control of the nutrients, vitamins, and minerals should be done according to the FSSAI or unavailability international guideline *Codex Alimentarius* may be followed.
10. Labeling should be done according to Food Safety and Standards (Packaging and Labeling) Regulations, 2011.
11. Manufacturing, packing, selling, offering for sale, marketing, distribution, or importing any package for use for special medical purposes are prohibited without following the regulation.
12. The following information should be there for special medical purposes foods.

13. The categorical words should be mentioned "FOODS FOR SPECIAL MEDICAL PURPOSE" at the close proximity of the name or brand name of the product.
14. The categorical statement in written boldly in a separate space should be written: "RECOMMENDED TO BE USED UNDER MEDICAL ADVICE ONLY".
15. The disease or particular disorder or medical condition for which the medication is to be used should be mentioned in the categorical statement "For the dietary management of (the condition)". Also, the validation of scientific data and claim should be proven by the clinical trial or studies in a larger population.
    e. The categorical statement of "NUTRITIONALLY COMPLETE" should be complying only with the ingredient of the foods is similar to complete.
    f. The expiry date should be mentioned.
    g. In the label, the targeted user group with the proper rationale of use should be mentioned. Also, the description and the properties of the foods should be mentioned.
    h. It should be mentioned on the label if the foods are to be used by a specific group.
    i. The use of nutritional ingredients should be mentioned with the rationale of including deleted, reduced, increased, or otherwise modified.
    j. The quantity of the nutrients should be expressed as recommended daily allowances.
    k. If applicable, the information of osmolality or osmolarity and/or on acid–base balance is required.
    l. All instructions, including the opening of the container, method of preparations, procedures of feeding, and storage conditions.
    m. The warning is also required that "NOT FOR PARENTERAL USE".
    n. The statement also required that the "Product should be stored out of the reach of children".
    o. Use of additives: As per regulated schedule VIII (c), (d), and (e) and Use for Foods for Special Medical Purpose.
    p. Contaminants, toxins, and residues: As per Food Safety and Standards (Contaminants, Toxins, and Residues) Regulations, 2011.
    q. Advertising Regulation: Approval is required from the FSSAI on a specific claim to serve in public.

10.7.8 Chapter VII (Foods containing Probiotic Ingredients)

1. Essential Composition: Only probiotic cultures of specific strain as mentioned in Schedule X of these regulations or any approved by the Food Authority.
2. The enlisting of the probiotic microorganism enlisted with scientific evaluation from time to time by the FSSAI.

3. The bacterial strain *Lactococcus* spp. (earlier known as *Streptococcus spp.*), *Lactobacillus* spp., and others used with substantiated probiotic action in the preparation of curd.
4. The microorganisms having probiotic properties with proper scientific validation study and should be approved by the Food Authority after proper scientific evaluation and enlisted in Schedule X.
5. Mention in labeling of foods containing Probiotic Ingredients
6. The labeling should be as per Food Safety and Standards (Packaging and Labeling) Regulations, 2011.
7. Manufacturing, packing, selling, offering for sale, marketing, distribution, and importing are prohibited without following the regulation.
8. The label should not attribute any disease-related claims, including preventing, treating, or curing human disease. Scientific validation is required about the structure, properties, and function of probiotics. It should be accompanied with a statement, "This product is not intended to diagnose, treat, cure or prevent any diseases".
9. Also, in each package following information should be accompanied:
   a. The categorical words "PROBIOTIC FOOD". The information of the strain like genus, species, strain designation including the International Culture Collection Number (International Culture Deposit Number), and the accession no of probiotics are required to be mentioned.
   b. After claiming and shelf life, there should be a minimum number of probiotic strains that are still alive.
   c. For the best physiological effects, the effective dose, serving vehicle, and length of treatment should be listed.
   d. Storage conditions with temperature designation and a "Best use before" label for the product after it has been opened.
   e. If there are known side effects, contraindications, or interactions with other medicines or foods, they should be written down in a warning. Any other precautions or warnings should also be written down.
   f. Also, the expiration date must be listed.
   g. Additives of Foods with Probiotic Ingredients: Additives should be in line with Schedule VIII (a) to VIII (e) of these regulations.
   h. Contaminants, residues, and toxins: According to the Food Safety and Standards (Contaminants, Toxins, and Residues) Regulations, 2011.
   i. Chapter VIII (Foods containing prebiotic ingredients)

10.7.9 Chapter IX (Specialty Foods containing ingredients based on Ayurveda, Unani, and Siddha, and traditional health systems of India)

1. Essential composition: Ingredients listed in Schedules V, VI, and VII that are used in Ayurveda, Siddha, and Unani systems for special foods or diets. This ingredient could come from a plant, an animal, a mineral, or a metal. It could be processed, unprocessed, or changed in a way that is acceptable

in a regular food that has been approved by the Food Authority. The form can be a liquid or syrup, a suspension, an elixir, granules, powders, tablets, capsules, or any other form that is acceptable.
2. No plant products other than those listed in Schedules V, VI, and VII are allowed.
3. Both the manufacturer and the importer must provide information about the product, including the herbal ingredient from Schedule V, VI, or VII that was used, how the quality was controlled, and how it was made.
4. The ingredients listed in Schedules V, VI, and VII are based on historical evidence recognized in the authoritative texts or monogram of the first schedule of Drugs and Cosmetics Act of 1940 and the Drugs and Cosmetic Rules of 1945.
5. In the first schedule, there are a list of 57 books about Ayurveda, 14 books about Unani, and 30 books about Siddha. Also included are the official monograms or formularies and Pharmacopoeias of the Ayurveda, Siddha, and Unani (ASU) systems.
6. The materials listed in Schedules V, VI, and VII can be processed according to the Ayurveda, Siddha, and Unani systems or any official books recognized by the government of India.
7. Schedules V, VI, and VII should be used to decide what the maximum dose should be, but the rule allows for more than one dosing schedule, and more than one dose can be given in a day.
8. The list is limited, and only those ingredients are used which are approval. The ingredients from countries other than India are not on the list. The foreign ingredients known to be used as food in other countries are required to have a safety profile for a minimum of ten years and a maximum of 30 years before being sent to another country.
9. Other than AYUSH, any ingredient that was not listed in Schedule V, VI, or VII had to be approved by the FSSAI before it could be used in food businesses. The scheduling of the ingredient required the right scientific data and validation, such as the amount of material used, quality control, test methods, scientific literature with scientific and technical information, safety and health benefits, and a reported human intervention study. The Food Authority has to check and review all the data on all the documents.

The following scientific and technical information is needed,

1. The quality of each raw material.
2. Information about the ingredients in the formula and how it is made.
3. Details about the packaging, such as the material used, how it is packaged, and how long it will last.
4. Controlling the quality of the finished food or health supplement and figuring out how to test it.
5. Information about how safe the drug is and how it works.
6. Information about human studies, if any.
7. Status of approval or regulation in other countries related to product information.

10.7.10 Chapter X (Novel Foods)

10.7.10.1 Approval Conditions

Pre-conditions implied that without the approval of the Food Authority cannot manufacture or sell or any trade business. Both the manufacturer and importer have required submitting the required documents, along with an application for approval.

10.7.10.2 Approval Process

1. Manufacturers or importers of new foods or people who make ingredients for new foods are required to apply to the food authority with all the details, such as a description of the product, the technology used to make the ingredients, and how the food is made, as well as the claim and statement reflected on the label and the required scientific data proving safety and effectiveness as follows.
    a. Common name of the new food.
    b. The details of the manufacturer or importer.
    c. A short summary of the new food.
    d. Product description.
    e. Contents.
    f. Details about the technology, such as how the foods are made and how they are preserved, packaged, and stored.
    g. Any proposed claims or statements that will be put on the label, along with enough scientific data to back them up.
    h. Information on how safe and effective the food is.
    i. Developing information.
    j. Analysis methodology.
    k. Shelf-life data.
    l. Undesirable effect.
    m. Information about foods that are different from the norm, if any.
    n. Details about directions of use and purpose.
    o. Status of the other country in terms of usage, regulatory approval, etc.
2. The book's literature or other relevant documents should show its process of preparation and confirming its uses as safe.
    a. Information about the maximum dose.
    b. The text that will be used on the labels.
    c. Information about the applicant's signatory.
    d. The new food also needs the following information.
    e. Chemical composition.
    f. Changes to the surface/chemistry of the surface.
    g. Particle size.
    h. Solubility.
    i. Digestibility.
    j. If any nanomaterials are added to the product.
    k. Specific claims.
    l. Labeling requirement of novel foods.

m. The mandatory requirements set out by the Food Safety and Standards (Packaging and Labeling) Regulations, 2011.
n. Specified claims should be made in accordance with the specific regulations or as updated by the Food Authority from time to time.
o. It is against the law to sell, make, package, or do any other kind of business in the name of "novel food" without the proper labels.

BIBLIOGRAPHY

1. Food Safety and Standards (Food Products Standards and Food Additives Regulation) 2011. Food Safety and Standards Authority of India. Available: www.fssai.gov.in/upload/uploadfiles/files/Compendium_Food_Additives_Regulations_08_09_2020-compressed.pdf
2. Food Safety and Standards (Food Products Standards and Food Additives) 2011. Food Safety and Standards Authority of India. Accessed: 21 March 2021. Available: www.fssai.gov.in/upload/uploadfiles/files/Food_Additives_Regulations.pdf
3. Gazette Notification on FSS (Food Product Standards and Food Additives) 2016. Ninth Amendment Regulation, Related to Standards for Proprietary Foods. Accessed: 19 October 2016.
4. Mohammed vs Union of India on 10 October 2014. Verdict by Hounrable Madras High Court W.P. No. 24999 of 2014. Accessed: 21 March 2021. Available: https://indiankanoon.org/doc/53359931
5. Swami Achyutanand Tirth & Ors vs Union of India & Ors on 5 August 2016. Verdict by Honorable Supreme Court of India WRIT Petition (C) NO. 159 OF 2012. Accessed: 21 March 2021. Available: https://indiankanoon.org/doc/75522969
6. Da vs Sushil Kumar & Ors on 21 August 2018. Patiala House Courts, New Delhi CC No. 49208/2016. Accessed: 21 March 2021. Available: https://indiankanoon.org/doc/149749963/

11 Herbs as Food vs. Drugs *Regulatory Framework*

11.1 INTRODUCTION

The historical and contemporary uses of herbs, emphasizing their roles as medicines, flavorings, and colorings. It also notes the importance of herbs in modern times as food supplements with significant nutritional value. Additionally, it mentions Dr. Stephen L. DeFelice as the first person to provide a definition of nutrition. Dr. Stephen L. DeFelice, the founder and chairman of the Foundation of Innovation Medicine (FIM), is credited with providing a definition of nutrition. Unfortunately, the specific definition is not provided in the statement. Nutrition typically refers to the process by which living organisms obtain and utilize the nutrients necessary for growth, maintenance, and metabolic functions. It encompasses the study of food, dietary patterns, and the physiological processes involved in the utilization of nutrients. Nutritional needs change depending on how the body works and on things like pregnancy, nursing, aging, and having acute or chronic diseases or other medical conditions. Sometimes, the line between claiming that foods were good for your body or that they helped protect against chronic disease or herbal drugs seems to be pixilated. Sometimes, food supplements and herbal or Ayurvedic drugs contain the same isolated herbal compounds. The problems with regulations come from the line between herbal drugs and nutraceuticals, and the fact that nutraceuticals are not defined by regulations. Also, the line between foods, nutraceuticals, functional foods, and dietary supplements is not clear.

In an age when the nutraceutical and functional food industries are becoming more global, laws are being put in place to make sure that regulations are the same in different countries. There is no doubt that nutraceuticals are a big part of the process of making new medicines, as long as strict quality controls are in place. As metabolic diseases like obesity, heart disease, and diabetes become more common around the world, people are turning to more natural ways to control them. Nutraceuticals are used to help people with long-term and degenerative diseases like Parkinson's and Alzheimer's. Diseases like coronary heart disease, diabetes, obesity, osteoporosis, and cancer can be prevented and helped by eating certain kinds of foods. Scientists found that beta-carotene and lycopene protect the skin from UV damage, that lutein and lycopene help protect the heart from damage, that lycopene is helpful for prostate cancer, and many other things. Other popular herbal and botanical products include Ginkgo biloba, saw palmetto, ginseng, St. John's wort, etc., all of which are said to help with chronic diseases like cancer, metabolic diseases like diabetes, heart diseases, high cholesterol, and so on.

Scientists have tried to figure out how to explain these properties by changing or adjusting biological processes, such as activating the antioxidant process, cell proliferation, regeneration, signal transduction, gene expression, mitochondrial pathways, etc. Also, nutritional research is now based on evidence from the molecular level.

DOI: 10.1201/9781003398431-11

Medical professionals say that there should be a line between nutraceuticals and diet supplements. Dietary supplements and herbal remedies or complementary or alternative medicine are more alike at the point where they meet, and the line between them should be there. The amount or dose to be taken, the standards, and the number of vitamins and minerals should be regularized. Also, advertising should say where the information came from and what it says for sure.

It is very important to follow the Indian regulatory framework for marketing of any products in India, taking into account that an herbal drug is listed as a nutraceutical, and that there should be a difference between "nutrition" and "pharmaceutical". Nutritional products should only be able to say that they are foods or food products that have health and medical benefits, such as preventing and treating disease. In the claim, there should not be anything about how nutraceuticals can help with health. Isolated, purified chemical entities isolated from food in very small amounts have pharmacological and therapeutic potentials considered as medicine, while nutraceuticals are consumed in quite large amounts and possess no isolated compounds claimed to have physiological benefits or provide protection against chronic disease.

Quality control is one of the most important parts of figuring out and measuring different nutraceuticals. The marker compound is needed for the manufacturing process because of its physiochemical properties, such as identity, purity, and biological properties. Combinations of nutraceuticals and herbs are entered with claims from ethnopharmacology that they are used in Indian medicine or other traditional practices. From ancient times to the present, our kitchen diet is full of nutrient-rich fruits and herbs, as well as various phytochemicals, vitamins, starches, minerals, and other nutritive values that keep us healthy, boost our immune systems, and help us avoid getting sick.

11.2 DIETARY SUPPLEMENTS

There is not much difference between "dietary supplements", and they are all used the same way. Vitamins, minerals, enzymes, and other substances can be used in the form of soft gelatin capsules or tablets. Dietary supplements can be used to boost the number of vitamins, minerals, or enzymes in the body through soft gelatin capsules. In India, dietary supplements cannot be regulated under the Drugs and Cosmetics Acts, 1940 or the purview of FSSAI, 2006. In the United States and India, dietary supplements should be taken by mouth and used for a month. Dietary supplements can include herbs, minerals, vitamins, other plants, and substances like amino acids, enzymes, glandular tissue, organ tissue, and metabolites. Dietary supplements can also be extracts or concentrates, and they can come in many different forms, such as tablets, capsules, soft gels, gelcaps, liquids, or powders. "These statements have not been evaluated by the Food and Drug Administration" must be written on the label. This product is not meant to diagnose, treat, cure, or prevent any disease".

On the Indian market, there are a number of herbal and non-herbal products—such as *Chyawanprash, Musli Pak, Ashwagandhadi Leh*, and cod liver oil—that are meant to be better sources of minerals, antioxidants, micronutrients, etc. Also, in common use, salt with iodine supplement is used to prevent goiter, folic acid supplement is used to prevent anemia, and wheat flour is used to prevent iron deficiency.

Special dietary supplements can also be taken in capsules, tablets, granules, and other unit forms like sachets of powder, etc., and have nutritional or physiological effects. Dietary supplements and nutraceuticals are not in the claim that they cure diseases, but they can be used to prevent diseases or make them better. Dietary supplements can be grouped by claim what chemicals content, or how they are used are as follows:

1. Probiotics.
2. Prebiotics.
3. Dietary Fiber.
4. Polyunsaturated fatty acids.
5. Polyphenols.
6. Antioxidant vitamins.
7. Spices.
8. Specially processed food supplements.

11.3 FUNCTIONAL FOODS

a. Functional foods are foods that have been added without changing their natural state, like vitamin-enriched apple oil, flour, and malt-based powder, as well as prebiotic or probiotic foods like yogurt. Functional foods are either high in nutrients or made through a process called nitrification. Both of these methods are used to prevent disease or make a special health claim, but functional foods share some qualities.
b. The foods should be present in their natural or natural form rather than a tablet, capsule, granules, or powder.
c. The foods should be able to be consumed every day in the form of food.
d. They should prevent or control disease and have specific physiological benefits or lower the risk of chronic disease.
e. At the same time, the regulatory body puts limits on the claims made about functional foods.
f. The number of certain nutrients should be mentioned at a certain level.
g. Functional claims should be made, and a description of how functional foods affect normal bodily processes and functions is needed.
h. The claim should only focus on the health benefits. In certain claim diseases, the restriction is implied not to be about a disease or an improvement in health.
i. All health claims about parts of the diet and a lowered risk of disease must be backed up by credible scientific evidence or logical reasoning, which the FSSAI reviews.

11.4 CLAIMS FOR FUNCTIONAL FOODS

Recently, it has been found that putting in a certain number of herbs leads to claims about the food's value, but some plants are also used in different functional foods with different claims. Common claims about the activities are discussed in what follows.

11.4.1 Antioxidant Properties

Most fruits and vegetables have antioxidants, and studies have shown that eating a of fruits and vegetables protects against a wide range of diseases and slows down the aging process. The idea of free radicals forming in the body is important for a number of reasons, such as the progression of cancer, heart disease, and metabolic diseases like diabetes and obesity. Free radicals are formed through like smoking, air pollution, and carcinogens in the environment. As a result of the fact that free radicals have to interact with DNA and the cell cycle, the normal physiological system is a bit different. Even though the body makes its own antioxidants and has other ways to protect itself, it is thought that antioxidants from the food we eat also play an important role.

11.4.2 Anti-Cancer Activity

It is not clear what causes cancer, but oxidants have definitely played a role. Most cells and DNA are damaged as cancer gets severe. Lemon, fennel, broccoli, and other foods can protect the DNA by stopping the neutralization of free radicals and boosting the enzymes. Also, different plant products help bind metal ions together and have antioxidant effects.

11.4.3 Anti-Microbial Activity

Several plants have anti-microbial effects, but there is not much clinical evidence to support this. Few plant parts are used as food ingredients to treat common stomach problems like diarrhea, upset stomach, etc.

Low-calorie foods and alternatives to sweating agents, such as the use of stevia, are common in foods like herbal tea, sugar-free sweets, and other modified foods that diabetic patients often eat, but a few clinical trials are available.

11.4.4 Anti-Inflammatory Effects

Components of the plant such as phenols, flavonoids, carnosol, and thymol, have anti-inflammatory properties. The plant products are used less in food than they are in tropical applications, and it is common for Indians to mix haldi with milk to treat inflammation. Chronic inflammation is indeed associated with various diseases, including conditions like irritable bowel syndrome (IBS), certain cancers, and neurodegenerative disorders like Alzheimer's disease. Inflammation is a complex biological response, and chronic inflammation can contribute to the progression of these diseases. It's important to note that chronic inflammation itself is not a symptom but rather a process that may be linked to various symptoms and diseases.

Flavonoids are a class of naturally occurring compounds found in plants, and they are known for their antioxidant properties. They are not typically added to foods like card (assuming you meant "card" and not a specific food or ingredient).

Foods rich in flavonoids include fruits, vegetables, tea, and red wine. Flavonoids have been studied for their potential health benefits, including their anti-inflammatory and antioxidant effects. However, the idea of adding flavonoids to make foods "safe to eat" might be misleading. Foods are generally safe to eat if they are prepared and handled properly, and the safety of a food product is usually not enhanced by adding flavonoids.

11.4.5 Brain Functions

The brain is one of the body's fully completed systems, and foods with glucose, antioxidants, and anti-inflammatory properties are good for brain function. Another important thing to remember is that the brain is not a highly permeable organ, and there is a strong, highly impermeable barrier system called the blood-brain barrier (BBB) system. Neurodegeneration is common in diseases like Alzheimer's, whereby damage to glial cells and nerve cells is caused by free radicals. St. John's wort, thyme, bramhi, shankpusphi, and a number of other herbs are used as foods.

11.4.6 Cardiovascular Function

There are many foods which can be used to lower cholesterol and raise trans fats. Foods with fiber, vegetable oil, fish oil, and food oil are used to lower cholesterol and triglycerides. Garlic powder, curry leaves, and coriander leaves are used to flavor foods and have been shown to lower cholesterol.

11.5 BOTANICAL SUPPLEMENTS TO FOODS

When botanicals are used in food, this is a very important and fascinating issue. Instead of food, plants are used in herbal medicine. A number of problems are sometimes seen, especially with this type of food, including the following.

1. Sometimes, the ingredients in food do not come from plants.
2. Clinical trials do not report on the safety, effectiveness, or side effects.
3. The plant extract does not have any nutritional value. For example, most phenolic compounds, such as anthocyanins and flavonoids, have antioxidant properties and protect cells from damage, but they do not have any nutritional value.
4. Plants' phytochemicals change with the seasons and also change in their roots, stems, leaves, flowers, and fruits.
5. It is challenging to make botanical supplements because plants are not the same and are mixed up.
6. Most botanicals do not have long-term safety studies, toxicity studies, or proper clinical trials. As an example, ephedrine was used in dietary supplements to give people more energy and help them lose weight, but pharmacologically, ephedrine causes anxiety, and too much of it can cause high blood pressure, irregular heartbeats, and death.
7. Most plants have a wide range of chemical parts, such as vitamins, terpenoids, flavonoids, phytoestrogens, glycosides, alkaloids, and carotenoids. Families like Alliaceae (garlic) and Brassicaceae (mustard) have unique phytochemicals. For example, the Alliaceae family has allyl sulfur compounds and the Brassicaceae family has glucosinolates.

8. Adding diluents or modifiers to the food can sometimes make it toxic. For example, "herbal tea" does not have a good amount of caffeine, but when it is made, guarana or cola nut are sometimes added, which raises the caffeine level. Also, phytochemicals like pyrrolizidine and different alkaloids increase toxicity.

11.6 COMPOUNDS CONTAINING PHENOLS

Phenolic compounds are especially important for antioxidant activities because they can get rid of free radicals and bind to metal ions. Flavonoids are an important part of phenolic compounds. They have antioxidant properties and other health-promoting properties, such as being anti-inflammatory, anti-microbial, and anti-cancer. The other important part is phenolic acid, which also fights free radicals. Flavonoids have a number of health benefits, such as protecting cells from cancer, preventing heart disease, stopping blood platelets from sticking together, etc.

11.7 CAROTENOIDS

Carotenoids have a skeleton made up of lipids with long chains of hydrocarbons that are symmetrically unsaturated and have double bonds that are linked together. This conjugated structure acts as an antioxidant by stopping singlet oxygen and peroxyl radicals. Carotenoids also give colors like yellow, orange, and red, and they can be found in a wide range of fruits, vegetables, algae, fungi, and bacteria. Mammals like us cannot make carotenoids on their own, so we have to eat them to meet our needs. Carotenoids come from both plants and animals, like the yellow egg yolk and plants like basil, cabbage, etc. Carotenoids are important to the human body because they are a precursor to vitamin A.

Most of the time, chlorophyll hides the colors of the carotenoids in plants. This protects the skin and eyes from damage, and it has antioxidant and cytoprotective effects.

11.8 ESSENTIAL OILS

Essential oil is made up of terpenoids, which are grouped by their degree of unsaturation. Fatty acids make up the backbone of all essential oils. They are made up of things like terpenes, alcohols, aldehydes, ketones, acids, esters, oxides, and so on. Essential oil is volatile and cannot mix with water. Many essential oils can kill bacteria and fight free radicals.

11.9 OTHER NUTRITIONAL SUPPLEMENTS

Phytoestrogen is a hormone that is found in some plants. Some of these plant products are added to foods with the claim that they are dietary supplements or special foods. Most of the time, this kind of food is used for sports foods, ergometric foods, and special anabolic activities. The phytoestrogen is found in cordyceps fungus, mung beans, alfalfa, chickpeas, and many other herbs that are used in food. No rules have been made yet about how to control foods that contain substances that mimic human hormones. The less information there is about toxicity, the less we know about the mechanism and how well it works in the clinically. Some foods—like St. John's wort (*Hypericum perforatum*), enriched foods, and *Tanacetum parthenium*—also contain melatonin. People say that melatonin directly reacts with radicals that cause oxidative stress, such as hydroxyl, peroxide, peroxynitrite anion, and hypochlorous anions, and that it is a super antioxidant.

Dehydroepiandrosterone (DHEA) is an anabolic steroid. It is used to control weight, build muscle, boost immunity, and slow down the aging process. Anabolic

TABLE 11.1 ROLE AND FUNCTION OF DIETARY

Dietary components	Safety	Functions
Vitamins		
Vitamins are chemical molecules and enzymatic co-factors. The presence of sunlight or other UV radiation changes inactive vitamin D (Cholecalciferol) found in plant products to active vitamin D (Cholecalciferol). Vitamin K is manufactured by intestinal bacteria. The richest source of vitamins A and E is food. Carotenoids are the precursor to vitamin A, and non-vitamin compounds such as lycopene have favorable health effects.	Vitamins are typically taken from meals that are safe and non-hazardous. Excess dosages of water-soluble vitamins are excreted in the urine, but the excess doses of fat-soluble vitamins are harmful because these vitamins are retained in adipose tissue. As per the Institute of Medicine (IOM)/ Food and Nutrition Board (FNB) recommendation, foods and dietary supplements are carefully prepared and have acceptable amounts or tolerated upper intake levels (UL) for nutrients of vitamin dosages. As with other water-soluble vitamins, such as vitamin C and diarrhea, vitamin B6 also causes peripheral neuropathy. Niacin is used in small dosages to prevent pellagra, whereas nearly 3 grams of nicotinic acid per day are necessary to lower cholesterol levels. Niacin has a maximum daily intake of 35 mg. Higher doses of fat-soluble vitamins, such as 20,000 international units (IU) per day of vitamin A, can cause spontaneous miscarriages and birth abnormalities. In addition, excessive vitamin D consumption can lead to hypercalcemia, calcium deposition in soft tissues, and severe heart and kidney damage.	Vitamins act as antioxidants and prevent cell damage or as biochemical process accelerators, including enzyme action and precursors.

TABLE 11.1 (*Continued*)

Dietary components	Safety	Functions
Minerals	All minerals added in accordance with the prescribed pharmacopoeias or monograms, such as Indian Pharmacopoeia (IP), US Pharmacopoeia (USP), or British Pharmacopoeia (BP). A high mineral concentration inhibits the absorption and accumulation in other organs. In addition to retarding the absorption of other minerals, a high iron level limits zinc and copper absorption. Calcium is added with bonemeal and dolomite for enhanced absorption.	Minerals play a crucial part in metabolic and physiological activities such as muscular contraction, regulation of water and acid–base balance, nerve impulse transmission, and catalysis of biochemical events. There are a total of 21 important minerals listed. Several minerals play crucial roles, including iron for hemoglobin, calcium and phosphorus for bone development, selenium for antioxidants and endothelium integrity, and trivalent chromium as an insulin cofactor. Other minerals, including copper, magnesium, and zinc, serve as co-factors for a few enzyme functions. The unit of minerals is referred to as adequate intake (AI).
Amino Acids	Humans are incapable of synthesizing the nine necessary amino acids found in meals. The direct positive effects of amino acid supplements have not been recorded in clinical studies. However, amino acid supplements have been administered to critically ill patients.	Dietary reference intakes (DRIs) or other dietary reference values for amino acids have not been established. The excess amino acid supplement is detrimental to human health, as evidenced by the added amino acids increasing protein reserves and causing poor calcium absorption. L-tryptophan supplement for the brain neurotransmitter serotonin, reported eosinophilia-myalgia syndrome (EMS) characterized by white blood cells and significant muscle pain, demonstrates the adverse clinical effects of amino acid supplementation. Therefore, amino acid supplementation must conform to the established monograph.

steroids do make muscles stronger, and the International Olympic Committee has banned these foods from sporting events. Dehydroepiandrosterone (DHEA) is a steroid hormone produced by the adrenal glands and is a precursor to other sex hormones like testosterone and estrogen. It's not typically found in significant amounts in plants. However, the mention of Dioscorea species, Microstylis wallichii, Asparagus racemosus, and other herbs might be related to the presence of compounds that are sometimes called plant sterols or phytosterols.

The specific content of DHEA in plants can vary widely depending on factors such as the plant species, the part of the plant used, the growing conditions, and the methods of extraction and analysis. It's important to note that the term "phytosterols" or "plant sterols" refers to a group of naturally occurring compounds in plants, and DHEA itself may not be present in these plants.

If there is a reference to DHEA levels in plants, it's crucial to verify the accuracy and context of the information. Additionally, the percentage range you provided (0–120%) seems unusual, as percentages typically range from 0% to 100%. It's possible that the numbers may be referring to a concentration relative to some baseline or standard.

Gamma-butyrolactone (GBL) is a chemical that is used industrially, for example, as a solvent. It can be converted into gamma-hydroxybutyric acid (GHB) in the body. GHB is a central nervous system depressant that is sometimes abused as a recreational drug. However, it is not an approved food additive, and its recreational use can have serious health consequences, including respiratory depression, coma, and even death.

Phytochemicals are compounds found in plants that are not essential for human life but have been recognized for their potential health benefits. Examples of well-known phytochemicals include flavonoids, carotenoids, and glucosinolates. The compounds you mentioned—androstenedione and creatine—are not typical phytochemicals. Androstenedione is a steroid hormone, and creatine is a nitrogenous organic acid found in some foods and synthesized in the body.

Regulation of Food Additives:

Androstenedione is a hormone precursor that was once available as a dietary supplement. However, due to concerns about its safety and potential for misuse in sports, it has been banned by many sports organizations. The U.S. Food and Drug Administration (FDA) has also taken action to restrict its use.

Creatine: Creatine is a naturally occurring substance found in certain foods and is also produced in the human body. It is commonly used as a dietary supplement, and its safety has been studied. The regulatory status of creatine can vary by country, and it's often classified as a food supplement rather than a food additive.

Rules and Regulations:

Phytochemicals and Other Compounds: The use of phytochemicals and other compounds in food is generally subject to regulations that vary by country. Regulatory bodies, such as the FDA in the United States or the European Food Safety Authority (EFSA) in the European Union, establish rules and guidelines for the use of these substances in food products.

Compliance: Food manufacturers are expected to comply with these regulations to ensure the safety and proper labeling of their products. The regulatory framework aims to protect consumer health and prevent misleading claims or unsafe practices. A few listed dietary components are as follows.

11.10 BEVERAGES

The common purposes and diverse characteristics of drinks, emphasizing their role in quenching thirst, providing immediate energy, and sometimes offering additional health benefits. Let's break down the key points:

Primary Purpose: Quenching Thirst: The primary function of most drinks is to quench thirst. Whether it's water, juice, tea, or other beverages, the fundamental goal is to provide hydration.

Added Health Benefits: Some drinks go beyond simple hydration and include added health benefits. This could involve the inclusion of vitamins, minerals, or other nutrients that contribute to an individual's well-being. These added elements may provide energy or contribute to overall health.

Immediate Energy: Certain drinks are formulated to provide a quick energy boost. This is often achieved through the inclusion of ingredients that deliver a rapid source of energy, such as sugars or carbohydrates.

Health Claims: Drink manufacturers often promote their products by making health claims. These claims may suggest that the beverages can prevent diseases, encourage good nutrition, and contribute to a healthy lifestyle. It's important for consumers to critically evaluate these claims and consider the overall nutritional content of the drinks.

Encouraging a Healthy Lifestyle: Some drink manufacturers position their products as part of a broader healthy lifestyle. This may involve marketing messages that emphasize the importance of hydration, nutrition, and overall well-being. In these categories are sports or energy drinks and fruit drinks with added vitamins, glucose, etc. On the Indian market, the following are the three types of drinks that are sold. But the way law enforcement agencies regulate is the same in every case.

1. Enhancement beverages: These drinks do not make any special medical claims. Instead, they are made to make the body work better without changing its pharmacological functions. Examples include energy drinks, sports drinks, glucose drinks, high protein supplements, etc.
2. Beverages for specific conditions: In this category, the drinks are said to have health benefits based on facts from medical science. The only way to use the ingredient is to mix it each other, like with antioxidants, mineral supplements, or vitamin supplements in common foods.
3. Foundation beverages: In this category, health benefits are claimed boosting the nutritional value of foods. This could be an alternation of the process or an alternation of the process variable. Sometimes this group of foods comes up in the purview of the medical acts. This group of foods includes macronutrient supplements, probiotic foods (like yogurt), and foods with added nutrients (fortified flour, soups, biscuits, etc.). Foods that are said to help children grow are mostly covered by either the baby foods or medical regulations.

11.11 FOODS FOR MEDICAL PURPOSES

The name is "medical foods", but these kinds of foods cannot be sold without a prescription from a registered doctor. Individuals consider medical foods as foods that need to be given by mouth or through a tube in special situations, like when a person has specific nutritional needs. DCGI or USFDA should approve these in India's legal system.

11.12 DIFFERENCES BETWEEN SUPPLEMENTS, DRUGS, AND FOOD ADDITIVES

There is a difference between drugs, dietary supplements, and food additives. In the definition of drugs, it is assumed that they are mainly used to diagnose, treat, prevent, or reduce the symptoms of diseases. Food additives are used to keep the food's nutrition, texture, and other qualities for a longer time. Dietary supplements are added to the diet to increase the total intake of a substance. Dietary supplements do not have to follow the rules for food additives because they do not contain food.

Approval of food additives is a rigorous process that takes into account safety and effectiveness. FSSAI looks at the additives and decides if they are safe and effective enough to go on the market. All of the drugs, including herbal drugs, are within the purview of Drugs Controller General of India (DCGI), whereby the safety and effectiveness of the product are of the utmost importance. After a successful clinical trial with individuals, a drug is approved based on its manufacturer's claims. Efficacy must be shown through adequate and well-controlled studies, including tests on humans, which show the drug will have the effect stated on its label.

Also, supplements to food claiming to be "nutritional supplements" or "health supplements" are both reviewed and approved by FSSAI.

11.13 LEGAL ASPECTS IN THE INDIAN SCENARIO

India's market for nutraceuticals and dietary supplements is expanding. Most nutraceuticals and dietary supplements are sold in India under the name "Fast Moving Healthcare Goods" (FMHG). In 2006, a law called the Food Safety and Standards Act (FSSA) went into effect. The goal of this act was to make sure that different laws worked together and that there was no confusion about mixing food with drugs. Another goal of the law is to make sure that food products are regulated in the right way in the context of nutraceutical dietary supplements and functional foods. The FSSA, 2006, are on par with laws in the United States, Europe, and other parts of the world.

FSSA, 2006 took effect throughout India on 5 August 2011. FSSAI has a clear purpose, it does its task well by authorizing rules on a regular basis and having full control over licensing, registering food businesses, packing, labeling, quality control, food product standards and additives, etc. Foreign foods are controlled by FSSAI laws that are harmonized and perform throughout India. This law looked at a specific of legislation and direction to authorities in India how to control the production, sale, and distribution of food while making sure it is safe and meets standards. Even though the law is in place in India, herbs are still used in a way that makes it difficult for people to distinguish between drugs and foods.

1. Section 3(b)(i) of the Drugs and Cosmetics Act of 1940 defines herbal drugs. Section 22 of the Drugs and Cosmetics Act of 1940 defines Ayurvedic, Siddha, and Unani drugs. So, the name "drug shadow" covers a wide area. In this wide shadow, nutraceuticals sometimes come in the form of tablets, capsules, or liquids that contain vitamins and minerals and make certain claims.
2. An empty gelatin capsule is considered a drug by the definition of a drug. This means that any product, including nutraceuticals, is also within the purview of drug regulations.
3. When making nutraceuticals, the additives, like coloring and binding agents, are not clearly listed.
4. Claims for nutraceuticals and dietary supplements as drug

11.14 THE PLACES WHERE NUTRACEUTICALS AND FOODS MEET ARE THE POINTS OF CONVERGENCE

Nutraceuticals and foods are fixed in a very narrow line, and each has its own domain, including content, claims, and quality control.

The content should be well written and should include the percentage or the amount. Nutraceuticals are taken in small amounts, while foods are eaten in large amounts. Nutraceuticals should be able to claim that they prevent disease or improve how the body works, but they should not be able to claim that they treat disease.

A case study is provided by *Glaxo India Ltd. vs. The State of Assam and others.* On 21 August 2002, the Honorable Gauhati High Court made a decision about how Glucon-C and Glucon-D are made and what they are made of—and sugar is the main ingredient, so Glucon-C and Glucon-D are not "food under a misbranded name". Instead, they are "proprietary foods".

In *Hindustan Lever Limited v. Food Inspector*, 2003 (Criminal Appeal No. 896 of 2003), the product name was "instant dairy whitener" which contained skimmed milk powder. The analysis report showed that the product did not meet the standard for skim milk. The case was brought under the Prevention of Food Adulteration Act, 1954, Sections 2(i–a) (a) and (m), 7(1), 16(1)(a)(i), and 17(1). The Honorable Supreme Court of India ruled that instant dairy whitener does not have any standards and that skimmed milk powder cannot be considered instant dairy whitener because it contains some skimmed milk, along with other ingredients. The standard can be set based on the important parts of each case.

11.15 FOODS AND HERBAL DRUGS

The Drugs and Cosmetics Act of 1940 regulates the drug, and most herbal drugs fall under clauses (a) and (h) of section 3 of the Drugs and Cosmetics Act (1940), which covers Ayurvedic, Siddha, and Unani drugs. Section 3 of this act says that a substance is a drug if it is: (i) used internally or externally to diagnose, treat, mitigate, or prevent any disease or disorder in humans or animals; (ii) used to change any physiological function or to decapitate microbes, parasites, or insects that caused the disease; or (iii) packaged in gelatin capsules or any other form of the unit dose form. Within the purview of the Drugs and Cosmetics Act of 1940 the term "not limited" is used to describe things that can be used both inside and outside the body.

The Drugs and Cosmetics Act covers a wide range of things, including health and nutritional supplements, special dietary uses, and functional foods.

Protein, vitamins, and minerals are all needed for the body to work properly, and the three parts have different effects on the body's functions, like blocking or activating receptors, enzymes, etc. But the food content should not be higher than the supplement in the form of a capsule or tablet. Interestingly, vitamin and mineral supplements are used to prevent and treat diseases. It is hard to know where the line between drugs and food supplements is. Along with the license for drugs and cosmetics, the FSSAI just recently also approved the vitamin and mineral supplement.

The case *Cadila Pharmaceuticals* Ltd. *vs. the State of Kerala and Others*, 2 April 2002 (AIR 2002 Ker 357), is about selling EC 350 (vitamin E and vitamin C capsules) and multi-vitamin capsules as "dietary supplements". The high court of Kerala said that the substances are included within the definition of "drug" under section 3(b), and that distribution, sale, and stocking should follow section 18 of the Drugs and Cosmetics Act.

11.16 DIETARY SUPPLEMENTS OR MEDICINE

Dietary supplements and medicine are the close points where the number of points went up. If you look at the two definitions, you can see that the main difference is that medicine is used to treat or improve a disease. This means that medicine is used very selectively for any disease in particular, while dietary supplements are used to keep people healthy and are taken in large quantities under normal circumstances. But some conditions—like malnutrition, pregnancy, and metabolic diseases like diabetes, heart disease, rheumatoid arthritis, etc.—where both medicine and dietary supplements are said to help, are bringing the two sides closer together. Using vitamin and mineral supplements in food is governed by the law, especially when they are used in special situations or to make certain claims. Chapters III and IV, Schedules D and K, of the Drugs and Cosmetics Act of 1940 made it clear that condensed milk or powdered milk (skimmed or malted), Farex, oats, lactose, and all other similar preparations that are fortified with vitamins and minerals, except those that could be used as drugs, are both food and drugs. Most of the time, things that have to do with food are covered by the under Prevention of Food Adulteration Act. However, the things previously listed are not only covered by food laws. Another interesting thing to think about is that this kind of food prevents diseases like vitamin or iron or other mineral deficiencies, and the drug does the same thing. Establishing a drug candidate is a difficult job, especially when it comes to herbal medicine, and it requires a lot of testing before and after it goes on the market. Also, the taxation is different for food and drugs.

In *Dr. Reddy's Laboratories Limited vs. the State of Kerala* on 12 June 2008 (RP No. 59 of 2006), the petitioner (Dr. Reddy's Laboratories Limited) was selling gamma linolenic acid (GLA) capsules, which can be used to treat diabetic neuropathy, and paying tax @12.5% rate. However, the assessing authority directed for taxation @20% equivalent to supplements/food preparations like Bournvita and Horlicks. The review petition was admitted by the Honourable High Court of Kerela and directed to assess whether GLA capsule falls under drug or dietary supplement and accordingly the tax needs to be decided.

11.17 CLAIMS IN LABELS

Claims of product elegancy are important, and claims should be limited to those as food only. The claims in the label should be as per the Indian regulatory guidelines, while selling in the Indian market requires India-specific labeling and packaging. Product-specific claims are also important and should not violate the Drugs and Cosmetics Act, 1940. Each component of the labels and how the product affects human healthcare should be noted. The drug candidate is required to a number of tests including pre-clinical studies, clinical studies, and safety studies, and the food composition should be not like a drug, but rather otherwise required to adhere to all regulatory steps. An example of litigation is *State of Bihar vs. Alkem Laboratories Limited, LPA* No. 865 of 2009, in which the Honorable Patna High Court directed the Director-General of Health Services, a government of India, under the Drugs and Cosmetics Act (1940), to report each component mentioned in the label are in the category of "food" or "drug".

BIBLIOGRAPHY

1. Gilbert, N. Regulations: Herbal medicine rule book. *Nature*, 2011; 480: S98–S99.
2. Chaudhary, A, Singh, N. Contribution of world health organization in the global acceptance of Ayurveda. *Journal of Ayurveda and Integrative Medicine*, 2011; 2.
3. Sahoo, N, Manchikanti, P, Dey, S. Herbal drugs: Standards and regulation. *Fitoterapia*, 2010; 81: 462–471.
4. Verma, N. Herbal medicines: Regulation and practice in Europe, United States and India. *International Journal of Herbal Medicine*, 2013; 1: 1–5.
5. Calixto, JB. Efficacy, safety, quality control, marketing and regulatory guidelines for herbal medicines (phytotherapeutic agents). *Brazilian Journal of Medical and Biological Research*, 2000; 33: 179–189.
6. World Health Organization. *WHO guidelines on good manufacturing practices (GMP) for herbal medicines*. Geneva: World Health Organisation, 2007.
7. Bansal, G, Suthar, N, Kaur, J, Jaon, A. 2016. Stability testing of herbal drugs: Challenges, regulatory compliance and perspectives. *Phytotherapy Research*, 2016; 30: 1046–1058.
8. Ramadoss, MSK, Koumaravelou, K. Regulatory compliance of herbal medicines—A review. *International Journal of Research in Pharmaceutical Sciences*, 2019; 10: 3127–3135.
9. Food safety and standards (Food products standards and food additives regulation). Food Safety and Standards Authority of India, 2011. www.fssai.gov.in/upload/uploadfiles/files/Compendium_Food_Additives_Regulations_08_09_2020-compressed.pdf

10. Benzi G, Ceci A. Herbal medicines in European regulation. *Pharmacol Res*. 1997 May;35(5):355–62. doi: 10.1006/phrs.1997.0132. PMID: 9299199.
11. *Codex Alimentarius Commission: Procedural manual*. Rome: World Health Organization Food and Agriculture Organization of the United Nations, Committee report, 2015.
12. Denham BE. Dietary supplements—regulatory issues and implications for public health. *JAMA*. 2011 Jul 27;306(4):428–9. doi: 10.1001/jama.2011.982. Epub 2011 Jul 5. PMID: 21730229.
13. EU Regulation (EC) No 1925/2006 of the European Parliament and of the Council of 20 December 2006 on the Addition of Vitamins and Minerals and of Certain other Substances to Foods. 2006; 50: 26–38.
14. FDA New Dietary Ingredients (NDI) Notification Process. 2019. (Accessed: 23 March 2021). www.fda.gov/food/dietary-supplements/new-dietary-ingredients-ndi-notification-process.
15. Sharma, S. Current status of herbal product: Regulatory overview. *Journal of Pharmacy and Bioallied Sciences*, 2015; 7: 293–296.

12 Regulatory Approval and the Commercialization Process of Herbals as Insecticides in India

12.1 INTRODUCTION

India is known for making generic pesticides. More than 60 technical-grade insecticides are made in India by more than 125 manufacturers, and the country has more than 500 insecticides pesticide formulators. People are becoming more interested in herbal insecticides as they learn more about organic foods. Herbal insecticides are useful for controlling pests and getting rid of insects that carry diseases. Herbal insecticides like neem, karanjin, annonins, rotenone, pyrethrins, nicotine, etc., are often used. Pathophysiology and how herbal insecticides work are well understood, but the likelihood of resistance is not well known. Herbal insecticides also work like common insecticides by affecting the nervous system, stopping larvae from growing (for example, by blocking the release of a hormone called "juvenile hormone"), and blocking enzymes that are needed for life. As for the problem of resistance, it is known that herbal insecticides are complex, so the chances of resistance are also very low. It is true that the analytical tools are not set up to find herbal insecticides, and there's a wide range scope of what people mean when they say something is organic or has little or no toxicity. Even now, export-quality tea, fruits, and vegetables are grown with the help of herbal insecticides. This is a big reason why herbal insecticides are in such high demand.

Herbal insecticides are not as easy to classify as chemical insecticides because they do not have clear structures. Instead, it is thought that all of the herb's parts work together to kill insects. But herbal insecticides can be put into different groups based on how they work, such as stomach poison, systemic poison, contact poison, residual poison, fumigants, repellants, and insect growth regulators. Herbal insecticides sometimes have two or three ways of killing insects. Most herbal insecticides are stomach poisons that kill insects when they eat them. On the other hand, some herbal insecticides kill insects when they come into direct contact with them, and most of these contain terpenoids. Most herbal insecticides with volatile oils are classified as fumigants and repellants because the volatile oils easily get into the respiratory system or because their strong smells keep insects away from the products. Except for nicotine at high doses, there are no herbal insecticides that leave a lasting effect. This type of insecticide keeps the lethal dose on the insect's surface, and it also kills other insects that are good for the environment, like ants, butterflies, etc. Very few or

DOI: 10.1201/9781003398431-12

no herbal insecticides have been identified as systemic insecticides. Because of this, herbal insecticides do not move from plants to insects. Herbal insecticides are used to stop the growth of insects by messing with their endocrine system, causing insects to change into something else, or stopping the synthesis or deposition of chitin. All herbal insecticides are controlled by the CIB and RC, which is in charge of the country's laws (Insecticide act 1968 and rules 1971).

12.2 A STEP IN THE HISTORY OF HOW THE CENTRAL INSECTICIDES BOARD AND REGISTRATION COMMITTEE CAME INTO FORCE

The disaster got worse when people in Kerala and Tamil Nadu died after eating imported wheat that contained insecticides. In 1958, the government of India set up a commission to look into the problem. From 1964–1967, the legislature and procedure were studied, and in 1968, the Insecticide Act was passed. This law covers the manufacture, sale, transport, import, and distribution of insecticides. In 1970, the Ministry of Agriculture took over the act's enforcement and set up the CIB and RC. The law went into effect on 1 August 1971, and is called the Insecticide Rules of 1971. All chemicals and formulas that stop, kill, or reduce insects, rodents, weeds, fungi, and other things that help plants grow but are not harmful to people are considered insecticides.

12.2.1 The Insecticides Act of 1968

The Insecticides Act of 1968 and the Insecticides Rules of 1971 are constitutional rules that control how insecticides can be used in India. The act has clear rules about how to register insecticides made in the country and those that come from other countries. The Insecticide Act also regulates the manufacture, sale, distribution, and transport of insecticides to make sure they are safe for people and the environment. According to the law, any chemical or substance that claims to kill insects must first be registered with the CIB and RC before it can be used or sold. In developing countries, insecticides are heavily regulated to protect health and improve registration skills. The Food and Agriculture Organization (FAO), the United Nations Environment Programme (UNEP), the US Agency for International Development (USAID), and the World Health Organization (WHO) all give advice to the legislative reforms about the risks, benefits, and help of the study of insecticides. They also help by providing training, analytical equipment, and analytical methods or tools. Section 4 of the Insecticide Act of 1968 gave the Ministry of Health and Family Welfare the power to keep an eye on the levels of insecticides in food. Also, the state government works with the central government to control insecticides. Usually, the state government is in charge of the following.

1. Measures to keep people and the environment safe during production, sale, storage, distribution, and use.
2. Use of the air and what it effects.

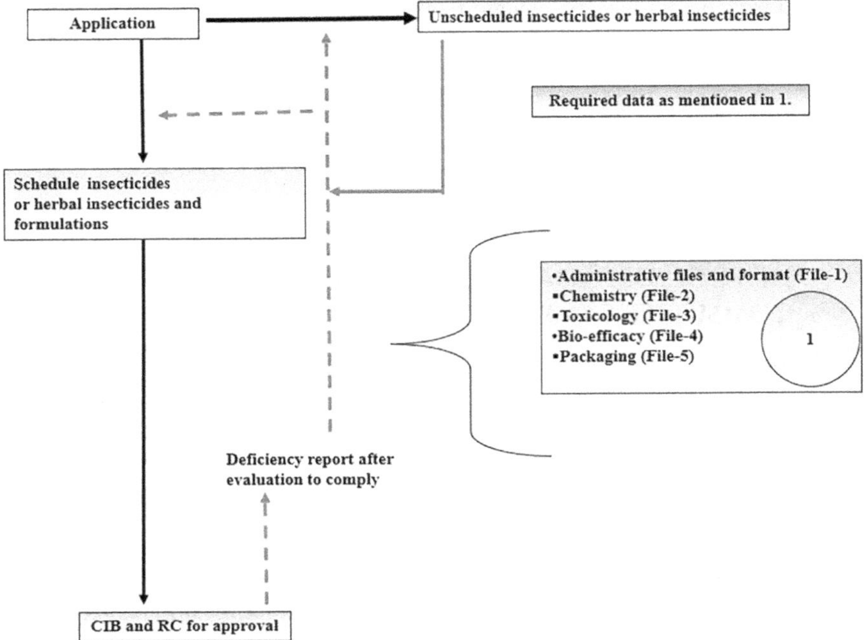

FIGURE 12.1 The process of registration of herbal insecticides in India.

3. Evaluation of shelf life.
4. Figuring out how much residue is acceptable.
5. Figuring out waiting periods.
6. Giving advice for colorization.

Recommendations are made for chemicals or substances on the schedule or the list of insecticides. Other tasks must be done as instructed by central government.

12.3 HERBAL PRODUCTS AS INSECTICIDES ARE IN THE PURVIEW OF THE INSECTICIDES ACT

Definition of "Insecticides": In Section 3(e) of the Insecticides Act of 1968, the term "insecticides" is defined to include:

(i) Any substance listed in the schedule.
(ii) Any other substances, such as fungicides and weedicides, that the central government may add to the schedule from time to time after consulting with the board and publishing a notice in the Official Gazette.
(iii) Any preparation that contains any one or more of the substances listed in (i). Schedule Inclusion: The substances listed in the schedule are considered insecticides under the Act. Additionally, the central government has the authority to

include other substances, such as fungicides and weedicides, in the schedule after consulting with the board and publishing a notice in the Official Gazette.

Expanding Scope: The Act provides flexibility to expand the scope of "insecticides" beyond the initial list by allowing the central government to include additional substances through a consultative process.

Importance of Proof: The statement suggests that the time it takes to prove that insecticides (pesticides) are used as claimed could influence whether they should be scheduled under the Insecticides Act of 1968. This implies that providing evidence or demonstrating the intended use of a substance as an insecticide may be a factor in determining its inclusion in the schedule. All registered insecticides are on a list, and the list is complete. So, if a new herbal insecticide has to be evaluated, it should be planned and covered by law.

12.4 ROLE OF CENTRAL INSECTICIDES BOARD AND REGISTRATION COMMITTEE

Under the section 4 of this 1968 act, Central Insecticides Board and Registration Committee (CIB and RC) this gives advice to the central and state governments on 15 pesticide-related issues, was established. Under the section 5 of the Insecticides Act, 1968, CIB and RC is formed with the power to approve insecticides and new formulations for use.

The CIB and RC looks at all the necessary data to make sure that insecticides work and are safe. Labels and leaflets are important for showing what kind of insecticide it is and how it can be used in agriculture or housed hold. On the label, it should mention what it is made of, what the active ingredient is, how much to use, what pests it is meant to kill, and any other safety tips. Household insecticides are different from agriculture insecticides. Under the Prevention of Food Adulteration (PFA) Act, 1954, the Ministry of Health and Family Welfare now sets the maximum residue level (MRL) values for insecticides in food and goods. This is done to make sure that people are safer.

12.5 THE INSECTICIDE ACT (1968) AND CHAPTERS

1. Chapter I ("Preliminary"): Short title, beginning, and definitions were included. The law goes into effect on 30th October 1971.
2. Chapter II ("Duties of the Board and Registration Committee"): In addition to the duties that the Act gives it, the board must also do the following.
3. Chapter III ("Registration of Insecticides and Method of Registration"): This chapter explains how insecticides are to be made, stored, and sold in India, and gives the committee the power to test the factory's "testing facility". The certificate will be given out in either Form II or Form II-A, or in a special format set by the inspection committee.
4. Chapter IV ("Granting Licenses or: Permission to Make Insecticides"): Form III or Form IV is used to get a manufacturing license, and 5000 Indian rupees should be deposited in the name of "Account officer, Directorate of the PP & S".

a. License agent: The state government put a notice in the Gazette and named a license officer who is responsible for following the law.
b. Grant of license: The insecticide described in the claims or the one that is set to be sold, stocked, or distributed, or that will be used as part of a large-scale pest control operation or for commercial purposes, will get a license from the license officer CIB & RC.
c. Licenses to make insecticides: Licensing of insecticides is done on Form III or Form IV with a prescribed fee of Rs. 2,000, and each additional manufacturing unit must apply for a separate license. Any changes to the manufacturing staff must be reported to the license officer. ISI mark is also needed to sell the insecticide, and after registering the insecticide, an application for the
d. Compliance with Indian Standards Institute (ISI) mark is needed. Forms VI and VII are used to renew a license, and the required fees are paid along with them. If an insecticide is to be stored, sold, or distributed from more than one place, it needs a separate license. Separation is needed for expired insecticides with names like "not for sale", "not fit to use", "not to be used", "not for use", and "not for manufacture".
e. Issuing cash memos and keeping records: All sales of insecticides should be recorded on Form XIII by the distributor, dealer, bulk consumer, or retailer, and all sales should be reported to the licensing officer once a month, including how much each person used. Form XIV, "Consolidated Statement of Used and Unused Insecticides", is sent back to the license officer within 15 days of each month. The license is then extended for two years.

5. Chapter V ("Packaging and Labeling"): It is limited or forbidden to sell, distribute, or transport insecticides that do not have the correct labels.
   a. Name of the manufacturer: The name of the person who is connected to the business should be given.
   b. Name of the insecticide: This should include the brand name, trademark, or any other common name, as well as the registration number.
   c. Name of active and other ingredients: According to the International Standard Organization (ISO), the common name of the chemical agents must be listed.
   d. Net content: The net content needs to be expressed in weight, measure, or quantity of units of activity rather than volume. The metric system for measuring volume and weight is also used.
   e. Batch number.
   f. Expiration date: The expiration date is important to know if the insecticide will keep working and be safe in the same way or in the same package.
   g. Antidote statement: if there's a chance of an accident, this must be declared. Most of the time, a common antidote is given, such as "If poisoning symptoms like nervousness, anxiety, tremor, convulsions, and allergic reactions show up, treat the symptoms" or "If accidentally

TABLE 12.1
Sub-Rule 3 as per the toxicity triangle

	Significance	Oral: LD_{50} mg/kg of Body Weight	Dermal: LD_{50} mg/kg of Body Weight	Inhalation: LD_{50} mg/l of Body Weight	Toxicity in the Organs	Toxicity Triangle
12.1.1	Highly toxic	≤ 50	≤ 200	≤ 0.2	Systemic toxicity observed seriously in oral, dermal, or inhalation exposure	POISON
12.1.2	Highly toxic	≤ 50	≤ 200	≤ 0.2	No systemic toxicity but severe contact toxicity in eye, skin, and respiratory system	DANGER
12.1.3	Moderately toxic	50–500	200–2,000	0.2–2.0	Moderately doses show systemic and/or contact toxicity	POISON
12.1.4	Slightly toxic	500–5,000	2,000–20,000	2–20	High doses show systemic and/or contact toxicity	CAUTION
12.1.5	Rarely toxic	≥ 5000	≥ 20,000	≥ 20	Slight or minimal systemic toxicity or contact toxicity	No symbol

swallowed, drink a lot of water, let yourself breathe in fresh air, and see a doctor right away". If the skin is infected, wash it with soap and a lot of water, or if the eyes are affected, flush them with a lot of water. If an antidote is known, it must be mentioned.

h. Label attachment: A label is stuck in the middle of the container so it cannot be easily taken off.
i. Label requirements: It should be in a prominent place and have a toxicity triangle set at an angle of 45° (in the shape of a diamond) as per Rule 4. At least one-sixth of the label's total face area should be a square.
j. The instruction in the label square is written in accordance with sub-rule 3.
k. Category I: Extremely toxic and marked with skull and crossbones and the word "POISON" in red ink on the triangle's top. "KEEP OUT OF REACH OF CHILDREN" or "IF SWALLOWED OR IF SYMPTOMS OF POISONING OCCUR, CALL PHYSICIAN IMMEDIATELY" should be the standard warning.
l. Category II: Chemicals in this category are considered to be highly toxic and should be labeled as "poison", and a warning should be placed outside of the triangle saying, "KEEP OUT OF REACH OF CHILDREN".
m. Category III: Chemicals in this category are labeled "DANGER", and the label must also say "KEEP OUT OF REACH OF CHILDREN" in a suitable place outside the triangle.
n. The chemicals in this category are "CAUTION" chemicals because they are only slightly dangerous.
o. As a result, the lower part of the triangle's color was given a name based on its toxicity.
p. The precautionary statement of sub-rules (3)–(5) of the label should say that the material is highly flammable or that it should be kept away from heat, open flames, or children.
q. Warnings should be printed in Hindi, English, and maybe one or two other local languages. Depending on where the packages are likely to be stored, sold, or given out, the local language should be chosen.
r. Labels should not say things like "safe", "non-poisonous", "harmless", or "non-injurious" that are not true, with or without a phrase like "when used as directed".

12.5.1 Manner of Leaflet to Be Contained in a Package

1. Common name of the insecticide as adopted by the ISO, and where such a name has not yet been adopted, such other name as may be approved by the CIB and RC may be used instead.
2. Instructions should include identification of the plant disease, insects, animals, or noxious weeds for which the insecticide is to be applied and adequate directions for use at the time of application.
3. The label of any insecticide's innermost container and the outermost covering that the container is packed must bear the following information, either printed or inscribed in permanent ink.

4. The manufacturer's name.
5. Name of insecticide.
6. Registration number of the insecticide.
7. Name of active and other ingredients.
8. Net content of volume.
9. Batch number.
10. Expiry date.
11. Statement of antidote.
12. The label should be prominently positioned and occupy no more than one-sixteenth of the total surface area of the label. It should feature a square shape, positioned at a 45-degree angle (resembling a diamond shape).
13. The square in question will be partitioned into two congruent triangles. The upper triangle will display the designated symbol and signal phrase as outlined in sub-rule (4), while the lower triangle will showcase the designated color as indicated in sub-rule (5).
14. The label and leaflets to be affixed or attached to the package containing insecticides shall be printed in Hindi, English, and in one or two regional languages to be stored or sell.

12.6 CHAPTER VI: INSECTICIDE ANALYSIS AND INSPECTORS

This chapter describes about the qualifications, job duties, and powers of the person who analyzes and inspects insecticides. An insecticide analyst usually calls for any analysis and/or proper examination of any insecticide that was sent or to whom the sample was referred. After the tests, the insecticide analyst sends a report to the CIB, RC, the insecticide inspector, or the place from where the samples came, as required by the act. The inspector of insecticides may give a report to the state government, and the report may be published or used in any way the government thinks is best for public interest.

The receiving and opening of a sized insecticide receipt is a technical process. Before the insecticide is opened for analysis, an insecticide analyst must compare the seals on the sample and the seals that are stuck on the insecticide. The specifications set by the Central Insecticide Board and Registration Committee CIB and RC are used to do the analysis. Under sub-rule (2) of the act, the report is made in triplicate on Form IX and given to the insecticide inspector.

12.7 CHAPTER VII: TRANSPORT AND STORAGE OF INSECTICIDES IN TRANSIT BY RAIL, ROAD, OR WATER

Packages containing insecticides are packed and transported by rail in accordance with the conditions specified in the rail tariff, issued by the Ministry of Railways. Direct contact with insecticide and foodstuffs or animal feeds is prohibited, and maximum precautions must be taken. This is required to ensure that no food materials are mixed with insecticide, and stores shall release after the thorough inspection of storage before being released to the consignees. If any of the insecticide is leaked, the countermeasure shall be taken to prevent the contamination of the environment, and the responsibility lies with the transport agency; accordingly, the state government is to be notified.

The warehouse or storage system is to be separate from the premises and stored in a locked separate room. The insecticide shall be not kept with the others article, and ideally insecticide is kept in the almirahs.

12.8 CHAPTER VIII: FACILITIES AND PREVENTIVE MEASURES FOR WORKERS INVOLVED IN THE MANUFACTURING OF INSECTICIDES

12.8.1 Protective Clothing

Appropriate clothing shall be provided to the persons involved in insecticide manufacturing, transport, distribution, or application. As per rule 40, along with the protective clothes, respiratory devices are to be provided. The protective clothes are to inhibit penetration by any insecticide during the handling of insecticide, and along with the protective clothes are provided rubber gloves reaching up to the forearms, dust-proof goggles, boots, and hat. The protective clothes are washable to remove the toxic elements after each use.

12.8.2 Respiratory Devices

To prevent toxic dust, vapors, or gases from getting into the lungs, the proper breathing equipment must be made available. Depending on the insecticide, full-face or half-face gas masks with canisters or respirators or gas masks are used. The most common types of respirators are chemical cartridge respirators, supplied-air respirators, and demand-flow respirators. During the manufacturing process, the air quality must be checked and the concentration of insecticides in the air must not be higher than what is allowed.

12.8.3 Aerial Spraying Operations

Ariel insecticide spraying is the operator's obligation, and it is normally done after notifying the competent authority for a minimum of 24 hours in advance to restrict access into a specific region. Only permitted insecticides and aerial distances are sprayed. The operator must receive particular training to perform first aid measurements in an emergency situation, as well as to dispose of insecticide packets, packages, and surplus insecticides to minimize environmental or water pollution.

12.8.4 Provisions of Antidotes for First Aid Measurement

The manufacturer or operator should keep sufficient quantities of antidotes and first aid medicines, first-aid tools, antidotes, equipment's, and injections by any exposure means. The potential routes of exposure include inhalation, dermal contact, ocular exposure, and ingestion. The manufacturers and distributors shall educate through the suitable training each operator for safety precautions and handling of the insecticides.

12.8.5 Disposal of Insecticide Containers

Used packages should not be used again. As a precaution, they should be broken up and buried far away from people's homes. Also, the packages must not be allowed to contaminate water or the environment.

12.9 CHAPTER IX: MISCELLANEOUS

According to the rule, insecticides can only be brought into the country through certain ports. Train stations in Firozpur Cantonment and Amritsar, the Ranaghat, Bongaon, and Mahiassan railway stations are part of a dedicated rail corridor. Chennai, Calcutta, Mumbai, Cochin, and Kandla are part of a dedicated sea corridor. Madras, Calcutta, Mumbai, Delhi, and Ahmedabad have been chosen to import insecticides through the dedicated air corridor.

12.10 PROCEDURE OF REGISTRATION OF HERBAL INSECTICIDES

The CIB and RC needs to register and license herbal insecticides before they can be imported or produced. The receipt of the application with the requested information and which confirms that the insecticide is registered. The RC may refuse to the grant of approval of registration and such cases there are provisions available to appeal to the committee.

12.11 DOCUMENTS FOR THE REGISTRATION OF PHYTO-CHEMICAL PESTICIDES/INSECTICIDES

The application requires the mention of the phytochemical's common and chemical names. Furthermore, data for chemistry, bio-efficacy, toxicity, packaging, and status of international registration are necessary. Many phytochemicals—such as *Pyrethrum, Azadirachtin, and Cymbopogon*—have already been certified by the CIB and RC.

12.11.1 Chemical Requirements

1. Many studies are required for the registration of new herbal compounds in order to establish the corresponding chemical entity. Plant parts are employed as insecticides or extracted as active chemicals. The manufacturing process, including a flow diagram, is required for identifying the chemicals. For example, azadirachtin is one of the major active constituents of neem extracts, and the specifications of the CIB and RC approved products describe that azadirachtin must be present in quantities of at least 1,500 ppm in neem-based formulations, 300 ppm in "kernel"-based formulations, and 300 ppm in "neem oil"-based formulations. If the insecticidal agent is not Azadirachtin, the application must specify the name, quality, and quantity. Important data such as chemical identity, physico-chemical

qualities, ingredient specifications, method of analysis, analytical test result, and shelf-life claim data are necessary. The following information, in addition to the registration of the new phytochemical, is necessary.

2. Source of the technical chemical.
3. Chemical composition in percentage of each gradient in the formulation.
4. Process of identification of the technical chemical.
5. Physico-chemical properties of adjuvants.
6. Technical bulletin of the final products.
7. Specification of the technical-grade active chemical or the active plant extract.
8. Methodology of analysis of the active constituents.
9. Analytical test report (ATR).
10. Identification and quantification of identifiable impurities.
11. Shelf-life claim.
12. Chemical equivalence establishment.
13. Details of the process of the manufacturing.
14. Raw materials information.
15. Source of the chemicals or active constituents.
16. Manufacturing process and its flow chart.
17. Chemical equation of the process.
18. Formula of the compounds.
19. Process of manufacture.
20. Effluent treatment process.
21. Standards of technical sample from the principals/authorized dealers for chemical verification.
22. Residue estimation methodology as per Bureau of Indian Standards (BIS) format.
23. Toxicity: Target toxicity is to be established to the rodents or birds with GLP compliance and in the guideline also directed for the toxicity assay to non-target or environment spices. The Organization for Economic Co-operation and Development (OECD) toxicity guideline is applicable for the testing. Tier 2 toxicity tests like carcinogenicity and mutagenicity assay are also directed. Along with that, documentation of effect on spray operators (health records) is also required. The following toxicity rest are mandatory for the registration.
24. Acute oral toxicity in two species.
25. Acute dermal toxicity.
26. Acute inhalation toxicity.
27. Primary skin irritation toxicity.
28. Irritation to mucous membrane toxicity.
29. Sub-acute oral toxicity.
30. Sub-acute dermal toxicity.
31. Toxicity to two bird species.
32. Toxicity to freshwater fish.
33. Toxicity to honeybees.
34. Medical data: risk assessment operators.
35. Human toxicity data information of any inclusive foreign countries.

12.11.2 Bio-Effectiveness

Bio-effectiveness is the assessment of efficacy in the laboratory and in the open field. It is to be established in three geo-climatic zones, and two years of data are required. Bio-effectiveness is also conclusive in terms of compatibility with other chemicals, and such data must be generated. In addition, bio-efficacy studies must explicitly state the aim of the manufacture, the direction of use, and the application process, as well as the timing, waiting intervals, and required equipment application. The following bio-efficacy tests are required for registration.

1. Bio-effectiveness in three geo-climatic zone for two years.
2. Phytotoxicity.
3. Persistence of the active chemical agents or degradation in soil.
4. Persistence of the active chemical agents in water or degradation in soil.
5. Persistence or degradation in soil and plants.
6. Compatibility with other chemicals.
7. Residues of the active chemical agents in plants.
8. Residues of the active chemical agents in soil.
9. Residue tolerance limits fixed by foreign countries.
10. Registration status of the other countries.

12.11.3 Packaging and Labeling

Packaging and labeling of insecticides are to be done according to the Insecticides Rules, 1971, which conforms the BIS format. The compatibility data, along with the container data, are to be studied. The following data are required.

12.11.3.1 Labels of Contents

1. Chemical composition or active content shall be mentioned in percentage basis (weight per weight [w/w], weight per volume [w/v], or volume per volume [v/v] basis).
2. Use or purpose of manufacturing.
3. Description of antidote in emergency.
4. Toxicity triangle as per lethal dose classification.
5. Cautionary or precaution if any, such as "avoid in bare skin" or "keep from children", etc.
6. Direction to application or uses.
7. Any restriction if any statement such as "toxic to honeybees or fish", etc.

12.11.3.2 Information Contained in Leaflets

1. Detailed chemical composition on leaflets accompanying small labels (up to 250 ml sized containers).
2. Introductory information about the pesticide.
3. Detailed directions about its use.
4. Application time.
5. Equipment details during application.

6. Waiting gap period after application.
7. Poisoning information and symptoms.
8. First aid measures in poisoning information.
   a. Specific and non-specific antidote and treatment.
   b. Restrictions, if any.
   c. Information about storage conditions.
   d. Instructions for the disposal of the container and/or the material after use.
   e. Type of packaging.
   f. Methodology of packaging.
   g. Details of the primary package as per BIS specifications.
   h. Details of the secondary package as per BIS specifications.
   i. Details of the transport package as per BIS specifications.
   j. Details of labeling.

BIBLIOGRAPHY

1. Shammi M., Sultana A., Hasan N., Rahaman MM., Islam SM., Doza BM., Uddin KM. (2020) Pesticide exposures towards health and environmental hazard in Bangladesh: A case study on farmers' perception. *Journal of the Saudi Society of Agricultural Sciences* 19: 161–173.
2. Isman MB. (2006) Botanical insecticides, deterrents and repellents in modern agriculture and an increasingly regulated world. *Annual Review of Entamology* 51: 45–66.
3. Central insecticides board and registration committee. ppqs.gov.in/divisions/central-insecticides-board-registration-committee (Accessed 14 February 2021).
4. Insecticides Rules. (1971) (GSR 1650, DT. 9–10–1971). www.bareactslive.com/ACA/ACT651.HTM (Accessed 14 February 2021).

13 Safety Studies or Pre-Clinical Toxicity of Modern Medicine

13.1 INTRODUCTION

India is a signatory to the World Trade Organization (WTO), and the quality and safety of Ayurvedic medicine are major considerations for its acceptability and marketing worldwide. India has a large and diverse range of protected species, and numerous herbal remedies have been shown to have special pharmacological properties. From a regulatory perspective, herbal medications have been largely standardized. Herbal medicines that have not been used previously or that have not been described in the literature and that make a replacement claim for currently available modern medication may be deemed new chemical entities and should adhere to the Drugs Controller General of India (DCGI) standards and pre-clinical safety studies as appropriate. The pre-clinical toxicity regulation requirement of DCGI and Ayurveda, Yoga and Naturopathy, Unani, Siddha and Homeopathy (AYUSH) is discussed in this chapter.

The primary and most significant concern in the development of any medicine is safety. Phase 0 or pre-clinical, or non-clinical research is the first step in determining a product's safety and further development studies. The probability of a hazardous reaction and the suggested cause are guaranteed during this stage of the drug development process. Before administering medications to humans or through any other method of drug administration, pre-clinical toxicology studies are a requirement in order to assess the adverse toxicity of the drugs.

Additionally, pre-clinical toxicology is used to comprehend how herbal medications may have negative side effects depending on dosage, quantity, interactions, and other physiologic factors. A pre-clinical study also outlined the maximum and minimum dosages, adverse reactions, target organ toxicity, and any side effects. Pre-clinical research assists professionals in choosing secure dose(s) during clinical investigations and afterwards in coming to a conclusion about therapeutic amounts. Safe dose selections are also generated from pre-clinical toxicity.

About 30 years ago, rules and pre-clinical toxicity assessments were modernized. Early in the 1950s, the regulatory approval process was fairly straightforward, and only a single acute rodent toxicity study was required. After the thalidomide incident in 1962, which shook the entire world, pre-clinical testing technique was examined, and new regulations were issued. Pre-clinical toxicity research design effectiveness depends on the dose selection, species, number, and age, and pharmacologist and toxicologist involvement in drawing conclusions from the pre-clinical study. Good laboratory practices (GLP) rules are currently being implemented in pre-clinical

DOI: 10.1201/9781003398431-13

toxicity research for harmonization and standardization. The GLP criteria for uniformity in pre-clinical evaluation technique were created and widely accepted by the USFDA (1979) and later by countries in the OECD (1981). In accordance with global harmonization, India agreed to abide by strict regulations and test criteria for the pre-clinical toxicological assessment of new pharmaceuticals.

Pre-clinical toxicities are more or less stated in the rules, and Investigational New Drug (IND) and New Drug Application (NDA) filings are more important in terms of the Indian government's regulatory clearance process. Pre-clinical toxicity studies are needed for the NDA and marketing approval processes. These studies must address chronic toxicity, phototoxicity, genotoxicity, carcinogenicity, and additional reproductive toxicity in F-1 and F-2 generations. Pre-clinical toxicity studies are fully documented for IND approval and include the acute, sub-acute, local, immunotoxicity, and genotoxicity. Studies on pharmacokinetics and toxicokinetics are crucial for determining a drug's "safe" and "toxic" exposure levels, as well as the maximum tolerated dose and the course of treatment. Pharmacokinetics also paints provide a picture of drug tissue depositions throughout organs and tissue distribution investigations. Absorption, Distribution, Metabolism, and Excretion (ADME) investigations are crucial for identifying metabolites and their advantageous or harmful effects. The development of various molecular strategies and biotechnological instruments hastened the completion of toxicology studies and improved the evaluation and forecasting of toxicities. The sensitivity of toxicogenomics is well recognized, and it offers new toxicity tests that go beyond the limitations of the present assays and detect toxicities using a clinical test that has been approved. The evaluation of pediatric medications differs significantly from conventional pre-clinical toxicity studies in that it uses young animals for animal toxicity testing.

Pre-clinical toxicity studies use sustained drug administration at various dose levels to detect long-term drug toxicity, such as tumor formation. A drug's unexpected and potentially fatal side effects, such as convulsions and arrhythmias, may occasionally or in very rare cases be found. The following pre-clinical tests are necessary.

1. The toxicity of new chemical entities, including acute toxicity and LD_{50} to determine the harmful dosage, median and, long-term toxicity, and maximum tolerated dose. Additionally, the potential negative effects on a number of particular organs—such as the brain, liver, kidneys, eyes, and genitalia—can be identified and anticipated appropriately.
2. Pre-clinical toxicity studies can be used as a guide when determining the risk or potential impact of a chemical in a human investigation.
3. To limit the possibility for toxicity, find alternative dose forms and administration modalities.

13.2 PRE-CLINICAL SIDE EFFECTS OF CONTEMPORARY MEDICINE AND A NEW CHEMICAL ENTITY

A majority of nations around the world accepts modern medicine. Modern medicine's top priorities including the safety issues and acknowledge the numerous reported negative reactions. Modern medicine is mostly plagued by three main problems. The manufacturing process's level of quality, homogeneity, and good manufacturing practice (GMP) compliance come first. The second problem concerns the need for clinical proof of safety and effectiveness, followed by the need for regulatory requirements and uniformity of the drug development process.

The term "new medications" or "new chemical entities" (NCEs) refers to drug candidates that are typically pure and well-characterized chemical compounds that have never been used in humans. The candidate is intended to be a "novel medicine", and DCGI is an authority of the regulatory approval process.

13.3 RECOMMENDATIONS FOR PRE-CLINICAL RESEARCH

Pre-clinical toxicity is managed globally in accordance with the following guidelines.

13.3.1 OECD Policies

WHO GCP Guidelines as detailed in the GLP Handbook of the World Health Organization. When it enforced to India, pre-clinical toxicity study is conducted in accordance with Schedule "Y" of the Drugs and Cosmetics (Eighth Amendment) Rules 1988. The regulation of dietary supplements falls under the category of foods, not pharmaceuticals. The USFDA does not conduct safety and efficacy testing on the majority of dietary supplements, unlike it does with pharmaceuticals. Dietary supplements are governed in the United States under the "Dietary Supplement Health and Education Act" (DSHEA) of 1994, in addition "Food Labeling Guide—April 2008" and "Guide to Nutrition Labeling and Education Act (NLEA) Requirements". Every dietary supplement must explicitly state its structure, its claim, and the mechanism of action, according to the DSHEA statutes. The manufacturer is in charge of making sure that these promises are accurate and genuine. Contamination, adulteration, and inconsistent dosing are possible concerns. The USFDA reviewed the reports of any adverse events or responses that were also tracked in dietary supplements.

13.4 GENERAL PRINCIPLES OF PRE-CLINICAL TOXICITY STUDIES

Toxicology studies should adhere to GLP standards or be accredited in order to assure excellent laboratory practices. All studies should be carried out in accordance with written protocols, with modifications (if any) being able to be verified retroactively. Studies should be carried out by appropriately trained and qualified staff using properly calibrated and standardized equipment of an adequate size and capacity. All management and laboratory responsibilities linked to these studies should be carried out in accordance with SOPs. The histology slides, as well as the approved protocol, raw data, draft report, and final report, should be kept for each investigation and kept archives for at least five years following the examination. To evaluate the link between the systemic exposures obtained in animals and the time course of the toxicity research, as well as non-clinical toxicity also known as toxicokinetic studies or specially designed studies should be carried out. The toxicokinetic study revealed the safety margin of exposure to specific medications and provides crucial information for safety pharmacology. The relevance of species selection, safety, and additional research were also underlined in this study.

13.5 SELECTION OF SUBJECTS IN PRE-CLINICAL TOXICITY

13.5.1 Rodents

Rats are the preferred rodent species for the majority of pre-clinical research using the suggested therapeutic administration method. Graded doses in four regimes are administered with one vehicle control, and the test chemicals should be dosed for ten days straight using a minimum of five animals of each gender. Using the maximum tolerable dose (MTD) as a guide, choose the greatest dose. Animals should be watched closely every day for indicators of intoxication (such as overall demeanor, activity, and behavior), as well as occasionally for body weight and other laboratory data such as gross inspection of viscera and a microscopic examination of affected organs are to generate.

13.5.2 Non-rodents

For the ascending phase MTD study, beagle dogs of both sexes are taken into account. The dose is logically increased by selecting 3–5 periods with the extrapolated efficient amount or MTD (whichever is smaller), and samples are taken for laboratory analysis every third day. When toxicity is evident, the dose should be reduced properly. More ten-day experiments are conducted, along with additional laboratory investigations at the well-tolerated dose intervals.

13.6 THE DURATION OF THE STUDIES

13.6.1 Repeated-Dose Toxicity Assessments Ranging from 14–28 Days

For the toxicity experiments, three doses are chosen from the range of highest dose level of visible hazardous reaction, mid-dose, and low dose. The dose chosen from the three-dose series based on the observed effective dose. Cage-side observations, food/water consumption, blood chemistry, bodyweight alterations, hematologic, and blood profiles are all observed as indicators of toxicity.

13.6.2 Toxicity Studies with Repeated Dosages over 90 Days

Three different dosages were chosen for the suggested clinical or therapeutic approach. Dosing should be conducted on a regular basis, following the pharmacologically prescribed route at three evaluated amount stages. It should have a "high-dose-reversal" group and a control group. General appearance, behavior, activity, body weight, food intake, blood clinical parameters, hematologic values, organ weights, urine analysis, and viscera and tissue autopsies are all analyzed. Half of the animals in the control and reversal groups are, whereas the other half are sacrified 28 days after the last treatment or when pathophysiological alterations are not evident. All of the studies are carried out in the manner as previously stated.

13.6.3 Study into Effects of Repeated Dosages over 180 Days

Repeated-dose toxicity studies for 180 days are recommended for long-term medications such as anti-diabetics, HMG-CoA reductase inhibitors, and so on. Toxicity testing is carried out at three different dose levels along the proposed clinical route. General appearance, behavior, activity, body weight, food intake, blood clinical parameters, hematologic values, organ weights, urine analysis, and a viscera and tissue autopsy are all considered.

13.7 SYSTEMIC TOXICITY TESTS

13.7.1 Effects of a Single Dose Studies

It is critical to determine the acute toxicity at the single-dosage level, as well as the minimum lethal dose (MLD) and maximum tolerated dose (MTD) in this study. Furthermore, in addition to the intravenous route in humans, one of the species has another route for systemic absorption may considered for comprasion. However, absorption is determined by the physiochemical characteristics of the medicine. The dose chosen was 10 times the usual human dose or 2 g/kg of body weight, whichever was greater. In this study, rats of either sexes are selected in a minimum 5 number, and they are observed for seven days before pre-dosing and 14 days after receiving an oral dose. Most of the time, four distinct doses are chosen in this type of trial, and if the drug is injected, the animals are given a single bolus dose or a continuous treatment over the course of 24 hours. General appearance, behavior, activity, body weight, food intake, blood clinical parameters, hematologic values, organ weights, urine analysis, and an autopsy of viscera and tissues are examples of parametric indicators. This study also looked for LD_{10} and LD_{50} values with 95% confidence limits, as well as any specific organ toxicity, if any. If LD^{50}s cannot be found, the appropriate explanation should be provided.

13.7.2 Dose-Ranging Research

This type of study is useful for assessing cytotoxic or anti-cancer medications to determine their toxicity to the target organ and determining the MTD for further studies. It is necessary to establish the linearity of response between the observed toxicity and the surface area of the employed species. Mice are typically utilized in this type of research. However, because some medications, such as anti-folates, do not affect rodents, non-rodent models are used to forecast toxicity.

13.7.3 Doses in Multiple Systemic Toxicity Research

This research is critical for a medicine with an unexpectedly long half-life, slow metabolism, organ deposition, inadequate elimination, and organ toxicity. The drug was administered to humans via its designated clinical route for seven days at three distinct doses, including the MTD, where toxicity was obvious, and a lesser toxic dose, where no toxicity was observed. The medium dose should be set logarithmically between the high and low doses. There should be no noticeable systemic harm at this dose. Furthermore, when single-dose tissue distribution studies, toxicity, and toxicokinetic data are available, this type of investigation is more persuasive.

The choice of species is a crucial aspect of this investigation, and the rats utilized in this study typically have metabolisms similar to humans. Two rodent species and one non-rodent species are chosen, and the indicators of toxicity are recorded after 14, 28, 90, or 180 days. The final dose-ranging study, however, will be performed at the therapeutic indication and scale of a proposed clinical trial. Long-term toxicity studies should monitor and record behavioral, physiological, biochemical, and microscopic findings. When administering a parent dose parenteral route, the injection site should be examined for toxicological responses and histology should be performed. When the medicine is administered to non-rodent animals, an electrocardiogram and fundus examination should be performed at the start and end of the study.

The anti-cancer medication trial is very important, and pre-clinical investigations are meant to match the therapeutic dose, schedule, study periods, and number of dosing of a proposed clinical trial. If Phase II, Phase III, or marketing approval is required, or if the medicine suggests a new mechanism of action, two rodent species are selected, and sometimes an additional one non-rodent species is suggested. A single dose is used to collect pharmacokinetic and tissue distribution data, but repeated doses are given if a single dose does not provide enough information.

13.8 TOXICOLOGY OF NOVEL COMPOUNDS

13.8.1 Men's Fertility Studies

Male fertility studies are considered on the dose selection discussed previously in this chapter, as well as the findings of 14- or 28-day toxicity tests. Rodents, usually rats, are used in this investigation. The results of a 14- or 28-day toxicity study are used to determine what dose to provide. Rats are usually the best choice, and the medicine should be administered in the same manner as in a clinical study. This research should take more than 28 days and no more than 70 days. The doses chosen are based on the one that demonstrated the lowest level of toxicity when evaluated at three different doses, as previously reported. After ensuring that the females are viable, male and females of the same species of rodent are matched in a 2:1 ratio for mating purposes. Male rats should continue to get their doses, while female rats should discontinue receiving them after ten days or when a vaginal plug forms, whichever occurs first. The fertility index can be calculated after 13 days of pregnancy. Each male rat's testes and epididymis should be weighed and histologically examined. The motility and morphology of the sperms from each epididymis should also be examined.

13.8.2 Female Reproduction Studies and Developmental Toxicity

The study's purpose is to determine how the drug affects women who are expecting to have offspring. Female reproductive toxicity research is classified into three kinds: teratogenicity, fertility, and perinatal investigations. The study employs two types of rodents. Teratogenicity parameters are studied using albino rabbits. When the rabbit does not respond well to antibiotic tests against gram-positive, anaerobic microbes and protozoa, the mouse is used instead.

13.8.3 A Study of Fertility in Female

Female fertility studies are conducted in the same manner as male fertility studies, with the dose determined by the median toxic dose (MTD) from prior systemic toxicity studies. The dose chosen should not have a negative impact on the health of the parent animals as a whole. Rats are employed the majority of the time in this investigation. Both the per dosage and control groups must have at least 15 males and 15 females. Males are dosed for 28 days, while females are dosed for 14 days. The dose schedules are the same before mating, during mating, and during pregnancy. It should be obvious that the administration route is intended for clinical usage.

Until the pups are mature enough to eat solid food, the dam should be given drugs, and the pups should be permitted to sip milk from the dam. Researchers must take note of how the animals mated, how far along they were in their pregnancies or giving birth, and how they behaved after giving birth during the experiment. Dams' body weight, food consumption, clinical indications of poisoning, and gross pathology (as well as the histology of damaged organs) should also be investigated. Intoxication, sex distribution, body weight, growth parameters, death (if any) gross examination, autopsy, and histology of the damaged organs are all closely monitored.

13.8.4 A Study of Fertility

The trials are carried out in stages, with the MTD derived from earlier systemic toxicity studies and the dose and animal species selected from one rodent (ideally a rat) and one other animal (rabbit). Treatment times for the mimic route of administration for clinical application are chosen during the organogenesis stage.

At each dose level, at least 20 pregnant rats (or mice) and 12 pregnant rabbits are utilized to examine the fetuses for abnormalities in the bones and organs. The body weight; the amount of food consumed; any changes in the uterus, ovaries, or uterine contents; the sites of implantation; the amount of corpora lutea; and resorptions are all symptoms of dam poisoning (if any). The number, gender, body length, weight, and any notable alterations or anomalies in the fetus's gross, visceral, or skeletal systems were also considered.

13.8.5 Perinatal Study

The study is significant because it will assist in determining whether there are any negative impacts on the development of a fetus in a pregnant or nursing mother. The method and dose should be the same or comparable to how it is given to humans. The study's protocols should comprise at least four groups, each having at least 15 dams, as well as a control group. The test drug regimen should be started on day 15 of pregnancy or in the third trimester, and it should be continued through breastfeeding and weaning. The dam is broken open at the end of the study to see if there are any abnormalities with the uterus, ovaries, or uterine contents, as well as the locations of implantation, amount of corpora lutea, and any resorptions.

The F1 generation study should be limited to at least four groups, all of which are treated with the same doses as the parents. Each group should consist of 15 males and 15 females. The development of sexual maturity, mating, pregnancy, birth, and nursing should all be investigated. Growth metrics should be monitored till weaning. The requirements should be the same as those outlined previously.

13.8.6 Genotoxicity

The drug should be evaluated for tumorigenic effects if it is going to be administered for a lengthy period of time (one year or more). Most anti-cancer drugs are cytotoxic; therefore, genotoxicity testing is not usually required before Phase I and Phase II studies, but it is required before Phase III trials begin.

A compound's genotoxicity or genetic damage can be assessed *in vivo* and *in vitro* using direct or indirect assays. Based on the results, hazards related to DNA damage and fixing can be identified and classified.

For determining mutagenicity, the following three approved test batteries are utilized.

1. Assessment of bacterial gene mutation.
2. Assessment of cytogenetic or chromosomal damage in mammalian cells, as well as an *in vitro* mouse lymphoma thymidine kinase test.
3. Assessment of chromosomal damage in rodent hematopoietic cells.

13.8.6.1 Gene mutation in bacteria

Gene mutation in bacteria is evaluated by Ames' Test or reverse mutation assay and tested in Salmonella species such as TA98, TA100, TA102, TA1535, TA97 or *Escherichia coli* WP2 uvrA (pKM101). A minimum of 5 log dose levels with "solvent" and "positive" controls should be employed by *in vitro* exposure with and without metabolic activation, S9 mix at least three replicates. As a control group, mutagenic agents such as 9-amino-acridine, 2-nitrofluorine, sodium azide, and mitomycin C are used. A 2.5-fold (or greater) increase in the number of revertants to spontaneous revertants is regarded positive.

Along with the conventional battery, DNA adducts, DNA strand breaks, DNA repair, or recombination should be considered as extra or further examination of genotoxicity. Alternative validated tests can be considered as standard battery substitutes, and rationale with appropriate scientific justification should be presented for doing a standard battery alternation. In the usual test battery, *Salmonella* and chromosomal aberrations (CA) were identified in cultured cells, whereas *in vivo* assays were identified as micronucleus assays (MNA) or CA in rodent bone marrow.

13.8.6.2 *In Vitro* Cytogenetic Assay

Most chemotherapeutic drugs require cytogenetic assays, which are performed using Chinese hamster ovary (CHO) or human lymphocyte cell lines. Accepted results indicate a 50% drop in cell populations or culture confluency to be evidence of cytotoxicity, and inhibitions of the mitotic index by more than 50% to be cytotoxic. A minimum of 3 log doses of S9 mix should be used for metabolic activation.

Cyclophosphamide with metabolic activation and mitomycin C without metabolic activation should be used as positive controls to ensure a reproducible, similar, and identifiable clastogenic impact. All studies should be done in triplicate, and the arbitration chromosomes in the metaphase should be considered for evaluation.

13.8.6.3 *In Vivo* Micronucleus Assay

In this experiment, a rodent species ideally a mouse is used, and the method of administration of the test chemical should be similar to that used in clinical trials. A minimum of five animals are used in each group for the experiment, and dosing is done in three levels, including "solvent" and "positive" groups. Positive controls such as mitomycin C or cyclophosphamide are employed as control groups, and dosing sequences should be performed on days 1 and 2 of the trial, with animals terminated six hours following the final injected dosage. The femoral bone's bone marrow should be isolated and flushed with fetal bovine serum. Samples are centrifuged and the bone marrow is pelleted, smeared on glass slides, and stained with Giemsa–May–Grünwald, and a minimum of 1,000 polychromatic erythrocytes with an increased number of micronuclei are employed as the evaluation criteria.

13.8.6.4 *In Vivo* Cytogenetic Assay

In this experiment, a rodent species, ideally a rat, is employed, and the method of administration of the test drug should be similar to that used in clinical practice. In this study, the results are examined in three dose levels, including "solvent" and "control", and each group should have at least five animals. This should be tested. Anti-cancer medications such as cyclophosphamide or colchicines are used as standard drugs, and the control drug should be administered day 1, the second dose should be given 22 hours after the first dose, and the animals should be sacrificed after two hours of the second day of administration. After the femoral bone was removed, it was flushed with hypotonic saline for 20 minutes before being pelleted and resuspended in Carnoy's solution. A low-speed centrifuge is required for pellet generation, and the pellets are sprayed on clean glass slides with a Pasteur pipette. The cells are stained with Giemsa stain, and an increased number of aberrations in meta phase chromosomes (minimum 100) is used as an evaluation criterion.

13.8.7 Carcinogenicity

Carcinogenicity studies are the most important, and the study's requirements are set by regulatory bodies. Carcinogenicity testing is essential for pharmaceuticals that are expected to be used for more than six months and drugs that are used often in an intermittent manner at the chronic or recurrent disease state. Furthermore, the drug's carcinogenetic characteristic defined the structure-activity connection, the chemical class of the pharmaceuticals, and the tissue retention of the parent drug or the metabolite of the product that causes pre-neoplastic lesion or tissue reactions. Carcinogenetic studies of drugs are sometimes considered based on their intended use, such as the condition of terminating patients in such disease circumstances when life expectancy is shorter than 2–3 years; in this instance, no carcinogenetic studies are required. Only the chance of secondary cancer is examined in cases when anti-cancer medications have a satisfactory therapeutic response and life expectancy is raised. Furthermore, the medications are effective for adjuvant therapy in tumor-free patients or for long-term use in non-cancer patients, and a carcinogenicity study is required in both circumstances.

With adequate rationale, the investigation is conducted in rodent species, preferably rats and occasionally mice. Rats are a suitable species since they do not have a very high or low prevalence of spontaneous tumors. Before optimizing the dose, three dose levels are tested. The MTD or highest dose should not be lethal, but rather sub-lethal, and the life span should not be lowered by greater than 10% of animals, although the minimum dosage ought to be equivalent to therapeutic dosage. The treatment time span is determined by a fraction of the life duration, which should be comparable to a fraction of the human life span. Dosing should be done every seven days, 24 months for rats and 18 months for mice. Each group and concurrent control group each have at least 50 animals of each sex. At the MTD or maximum dose level, at least 20 animals in each sex are recruited, whereas at the lowest dose level, at least ten animals in each sex are recruited. Furthermore, this study should be planned to thoroughly outline benign and malignant tumor development, time of identification, site size, histological typing, and so on. Body weight, food intake, blood clinical parameters, hematologic values, organ weights, urine analysis, and an autopsy of viscera and tissues are all taken into account in the parametric analysis.

Additional tests, such as short-term bioassays and neonatal mouse assays, are occasionally used to conduct the experiment in transgenic animals, and microscopic alterations should be seen at the time of autopsy. Carcinogenicity data are not required for large-scale clinical trials unless there is a categorical concern for the patient populations.

13.9 LOCAL TOXICITY

When the drug is intended for usage other than the oral route, the local toxicity is crucial. The majority of these dosage forms have limitations, such as anatomical complexity or local application, or drug bioavailability cannot be enhanced due to solubility, pH, or tonicity. These drugs are to be administered to an appropriate site (e.g., skin or vaginal mucous membrane) where the desired local effects are predicted in a suitable species. If the medicine is absorbed locally, system toxicity data are required. The study is also conducted with a three-dose level that includes the untreated and/or vehicle control, and preferably two species are employed with increasing group size as treatment duration increases.

13.10 DERMAL TOXICITY STUDY

Dermal toxicity studies are required if the formulation or medicine is planned to be used through the dermal route, such as ointment, cream, or lotions. The rodent species rabbit and rat are utilized to assess toxicity, whereas the guinea pig is employed for skin sensitization testing. The test material can be applied directly or using porous gauze to shaved skin at a minimum of 10% of total body surface area. The study term ranges from 7–90 days, depending on the clinical use and duration of use. Skin irritation was clearly visible during the observation, and the recovery group should be included in the subsequent repeated-dose trial. Other local symptoms—such as erythema, swelling, and eschar formation—are also noticed and validated by histopathological investigation of application sites.

13.11 PHOTOALLERGY OR DERMAL PHOTOTOXICITY

The drugs are treatments for metabolic disorders which are causing the changing of color of skin by photoreactions like leucoderma. Armstrong/Harber Test are ideal methods and tested in the guinea pig. Initial tests are carried out with eight animals to evaluate the eight concentrations. The drug is taken in the patch and applied to guinea pigs with and without UV exposure (10 J/cm^2). The highest non-irritant dose should be determined after application at 24 and 48 hours and should be used to ascertain the dose. After the obtained preliminary results, the main test is initiated by using ten test animals and five controls. The selected dose which determined in the initial study is applied in a patch of about 0.3 ml and keep on the tested animals for 2 hours, 15 minutes, followed by 10 J/cm^2 of UV exposure. The study should be repeated on the days 0, 2, 4, 7, 9, and 11 of the test. The test concentration should be the same and challenged between days 20–24 with a similar two-hour application followed by exposure to 10 J/cm^2 of UV light. After 24 and 48 hours after the challenge, the local signs are examined and graded including erythema, edema, and eschar formation. Musk ambrette or psoralen can be considered as the control agents.

13.12 VAGINAL TOXICITY TEST

Topical dosage, such as ointment or cream, is meant for application on soft skin, and in this instance, a vaginal toxicity study in rabbits or dogs is required. The test chemical should be administered to the vaginal mucosa, and 6–10 animals per dose group are assigned. The dosage was changed numerous times to the original clinical dose of humans, and the study's observation periods ranged from 7–30 days. The swelling, closure of introitus, and histology of the vaginal wall are rated, and conclusions are drawn.

13.13 RECTAL TOLERANCE TEST

Local administration or loading of higher concentrations via the rectal route is preferred over the oral route. To manage extreme fevers, a high dose and numerous doses of paracetamol might be provided. The test is carried out in rabbits or rats at several folds of the drug concentration and/or volume of clinical usage to humans. The test chemical should be applied to six to ten animals per dose group once or several times daily per rectally to determine the multiples of daily human dose, and the study's observation periods are limited to a minimum of seven days and a maximum of 30 days. The clinical indicators of pain are sliding on the backside, blood and/or mucus in feces, pathological alterations of the anal region/sphincter, gross and (if necessary) microscopic evaluation of the rectal mucosal are all graded.

13.14 OCULAR TOXICITY STUDIES

Ocular toxicity is to be tested by topically applying local infection of the eye or items intended for ocular instillation. The animals with a sufficiently large conjunctival sac are chosen for the trials, and the albino rabbit is the recommended species. The exposure length should be similar to clinical applications in the repeated dose, and recuperation intervals of up to 90 days are required. Initially, tests should be carried out to determine the exposure concentration for the repeating dose. The therapeutic index or safety margin is calculated at two doses. One eye should be utilized for testing and the other for control in acute toxicity testing. Slit-lamp examination of the cornea, iris, and aqueous humor is recommended, while fluorescent dyes (sodium fluorescein, 0.25–1.0%) should be used to detect deficiencies in the surface epithelium of the cornea and conjunctiva. Tonometers are sometimes used to determine ocular tension and histopathological inspection of eyeballs fixed in Davidson's or Zenker's fluid for additional tests.

13.15 INHALATION TOXICITY STUDIES

The majority of anti-asthmatic medicines or bronchodilators are administered in aerosol form for direct effect on respiratory organs. The final, marketable formulation is tested on one rodent and one non-rodent species. The design of the trials which include acute, sub-acute, and chronic toxicity tests is geared for clinical usage in humans. For the evaluation of gases or vapors, whole-body plethysmography or whole-body exposure chambers are employed, whereas aerosols are to be delivered via nose only. The limit test of the inhalation test is 5 mg/l, and the dose is changed in various folds in accordance with human clinical use. The animals are exposed in three groups, each with a control (plus a vehicle control, if necessary), and the period of exposure is set at a maximum of six hours per day, five days per week, with clinical use modified accordingly. Food and water should be removed from the experimental animals during the experiment and exposure to test material. Temperature, humidity, and flow rate of the exposure chamber should be monitored during tests, and the particle size range should be in 4 microns (particularly for aerosols) and not less than 25%. The impacts of respiratory rate including inspiration and expatriation, bronchial lavage fluid conditions, and histological analysis of respiratory passageways and lung tissue—should be included in the experiment. The safety margin or therapeutic index can be determined using the standard parameters of systemic toxicity assessments.

13.16 GUINEA PIG MAXIMIZATION TEST

This method involves conducting tests in two steps to establish the maximum non-irritant and minimum irritant doses. The preliminary study will include two components. The intradermal method should be used in a batch of four male and four female animals, with two of each sex receiving Freund's adjuvant and the least irritating dose calculated. In a similar technique, the minimal irritating dose for a challenge should be determined in two males and two females.

The main study begins with a minimum of six male and six female animals per group, with one test and one control group included. The test is sometimes performed with more than one control group. In the primary trial, intradermal induction (day 1) is followed by a topical challenge on day 21. If no results are obtained, a re-challenge is issued 7–30 days following the initial challenge. Erythema and edema development in the skin are assessed and graded. Other physiological monitoring, such as blood pressure and electrocardiograph, are also assessed.

13.17 LOCAL LYMPH NODE ASSAY

Typically, 3H-thymidine or bromodeoxyuridine (BrdU) incorporation assays are performed. In this study, a minimum of six mice in each group, either male or female, are recruited, and the test medication is administered to the ear skin for three days in a row. After the treatment, on the fifth day of dosage, 3H-thymidine or bromodeoxyuridine (BrdU) is injected, and after the fifth hour, the draining auricular lymph nodes are to be dissected to evaluate the results.

BIBLIOGRAPHY

1. Wu KM, Dou J, Ghantous H, Chen S, Bigger A, Birnkrant D. Current regulatory perspectives on genotoxicity testing for botanical drug product development in the USA. *Regulatory Toxicology and Pharmacology* 2010; 56(1): 1–3.
2. Drugs and cosmetics (first and second amendment) rules. Ministry of Health and Family Welfare, Govt. of India, 2008.
3. WHO. *Guidelines for the assessment of herbal medicines, programme of traditional medicines*. World Health Organization, Geneva, 1991.
4. WHO. *WHO guidelines on good agricultural and collection practices for medicinal plants*. World Health Organization, Geneva, 2003.
5. Mosihuzzaman M, Choudhary MI. Protocols on safety, efficacy, standardization, and documentation of herbal medicine (IUPAC technical report). *Pure and Applied Chemistry* 2008; 80(10): 2195–2230.
6. UCSF IACUC POLICY– Laboratory Housing and Study Areas for Research Animals. Adopted on May 2001, Revised December 2003.
7. Schedule "Y" in Drugs and Cosmetics (Eighth Amendment) Rules. Ministry of Health and Family Welfare, Government of India, 1988.
8. ICH S5(R2) Guideline: Detection of Toxicity to Reproduction for Medicinal Products and Toxicity to Male Fertility, June 1993.
9. ICH S1A Guideline: Guideline on the Need for Carcinogenicity Studies for Pharmaceuticals, November 1995.
10. ICH M3(R2) Guideline: Guidance on Non-clinical Safety Studies for the Conduct of Human Clinical Trials and Marketing Authorization for Pharmaceuticals, June 2009.
11. ICH S9 Guideline: Non-clinical Evaluation for Anticancer Pharmaceuticals, November 2008.
12. Abhishek, M., Rubal, S., Rupa, J., Bikash, M. (2023). Toxicology of Pharmaceutical Products During Drug Development. In: Mohanan, P.V., Kappalli, S. (eds) *Biomedical Applications and Toxicity of Nanomaterials*. Springer, Singapore. https://doi.org/10.1007/978-981-19-7834-0_7.
13. Müller L, Mauthe RJ, Riley CM, Andino MM, Antonis DD, Beels C, DeGeorge J, De Knaep AG, Ellison D, Fagerland JA, Frank R, Fritschel B, Galloway S, Harpur E, Humfrey CD, Jacks AS, Jagota N, Mackinnon J, Mohan G, Ness DK, O'Donovan MR, Smith MD, Vudathala G, Yotti L. A rationale for determining, testing, and controlling specific impurities in pharmaceuticals that possess potential for genotoxicity. *Regulatory Toxicology and Pharmacology* 2006; 44: 198–211.

14 Clinical Trial Procedures Including Modern Medicine and Ayurvedic Therapies

14.1 INTRODUCTION

India is a great nation with more than 1 billion people and considered the largest citizenry in the world. India is an authoritative name for itself in the all-embracing clinical trial as an adopted destination for companies to conduct clinical trials. It is a great challenge for both the government authorities and the companies to actualize a contrast between ethics and business. The Indian government has played a successful role by reforming and implanting standardized research protocols and successful trials are performed by pharmaceutical companies and clinical research organizations. Furthermore, an added acceptance of good clinical practice (GCP) requirements and a stronger admiration for all-embracing acceptability of research are being performed in India with human subjects and accepted by Indian clinicians in accordance with clinical trials. Investigators are inquisitive to implant clinical trials that accede to GCP and are also accommodating to working within the constraints of required protocols.

In 1995, India became an affiliate of the World Trade Organization (WTO) and agreed to follow its product patent regime in 2005. The pharmaceutical industry entered India by taking the safeguard and right of the patent act. After alignment with the WTO, the pharmaceutical sector of India drastically increased its growth and increased stakes of multinational companies in Indian operations. India is considered a major contributor in clinical trials after the WTO agreements and most multinationals see it as a favorable destination for the clinical trial. India is a true country in the respect of using English language in official documents, a wide spectrum of disease, accomplished with talent and intellectual environment, stable government, stable economy, and moreover the support from the administration and government. International consulting agency Ernst and Young and the Federation of Indian Chambers of Commerce and Industry disclosed that the Indian population participates more than 7% of Phase III and 3.2% of Phase II trials of the global. After permitting contract research organizations (CRO), the outsourcing of clinical trials has increased tremendously up to 30% and can be considered the most robust in the global market and proposed valuation of the market 2022 to about, $ 2.1 billion. At the same time, a big challenge is ahead with the government on certain issues including the rights and safety of the volunteers or patients and the regulatory

DOI: 10.1201/9781003398431-14

framework, and how to protect the ethical issues of the patients. It is true that in the judicial framework of the Indian government, the clinical investigator takes the major responsibilities of any ethical concurrence in a successive trial, and it is also true that the success or failure of any trial depends on the investigators and how the assigned trial acknowledges the responsibility and the commitment to the ethical conduct of trials. It cannot be denied that in a tremendous amount of increasing trials, the clouds of a threat to society are coming because of concerns about patient rights and safety. The government is working with the responsibility to look into the compliance of the laid rules in the trial, unethical trials, infrastructure, unethical practices, and training. Recently, a number of defaulters were prosecuted concerning the trial. Also, the government is committed to looking into the organization of Instituted ethical committee (IEC) data quality, infrastructure, involvement of finance, patient insurance, and other factors. However, all such barriers are common in all developing countries, and how these are addressed by authorities is the biggest challenge.

The Central Drugs Standard Control Organization (CDSCO) under the Ministry of Health and Family Welfare establishes standards and norms, controlling regulatory measures for drugs, diagnostics, and devices. The organization is also responsible for authorization of new drugs, and the amended acts and rules from time to time. CDSCO is also actively involved in standardizing clinical research in India and introducing safer drugs to the Indian market. Drugs and Cosmetics Rules of 1945 enforced the Schedule Y, where the schedule that was engraved with it outlined the standards, requirements, and rules for clinical trials in India.

Understanding the importance of streamlining the clinical trial and the ethical issue, the Indian Council of Medical Research (ICMR) came out with draft guidelines for clinical trials involving human subjects. The first drafted guideline prepared under the chairmanship of Honorable Justice (Retired) M.N. Venkatachaliah in 1996, and different subcommittees were formed to considering the specific areas of medical research including epidemiological research, organ transplantation, products to be used in human clinical investigations, human genetics, etc. The drafted guideline was published in 2000 and the last amendment was made in 2006 for the issue of using humans in biomedical research. The CDSCO released additional guidelines in 2001 on Indian GCP. Both guidelines are appreciable for the easily written in Indian perspective to consider all the international committee recommendation including those at Helsinki, Finland, June 1964 and amended by the 29th WMA General Assembly, Tokyo, Japan, October 1975; 35th WMA General Assembly, Venice, Italy, October 1983; 41st WMA General Assembly, Hong Kong, September 1989; 48th WMA General Assembly, Somerset West, Republic of South Africa, October 1996; the 52nd WMA General Assembly, Edinburgh, Scotland, October 2000; and Ethical Principles for Medical Research Involving Human Subjects - Helsinki guideline.

Implementation of Schedule Y of Drugs and Cosmetics Act in 1988 and regulatory guidelines for the clinical trial was established, which forced compulsory permission for any clinical trial. Manufacturers are required to register new drug after clinical trial Phase III and for the registration of a new drug, which leads to the growing pharmaceutical market in generic medicine. Despite the increased introduction of foreign pharmaceutical companies in the Indian market, the intellectual property and strict patent rules were kept according to the interest of the stakeholders and the government to introduce many changes including ethical and regulatory guidelines.

It was realized that to keep pace in the global scenario, Schedule Y was required to change, and revised Schedule Y was implemented in 2005. The revised Schedule Y is more rational and the sponsor companies are prevented from designing clinical trials; accordingly, the research comes out. The designs of the trial are well written. The amendment also strongly regulated the volunteers' rights including consent, IEC approval, and reporting of serious adverse events. The amendment of Schedule Y also leads to good clinical practice (GCP), as per global acceptance. The responsibilities also imposed on the IEC, investigator, and sponsor to maintain the auditable documents of a clinical trial.

14.2 LEGAL FRAMEWORK OF CLINICAL TRIALS

India is considered a final destination of clinical trials because of its large population, with a well-trained English-speaking workforce and significantly low costs compared to other parts of the world. For any new drug or new designed chemical entity, the clinical trial started with permission from the Drug Controller General of India (DCGI) after submitting the requisite documents like informed consent forms, investigator details, Ethical Committee permission, and nowadays the insurance and compensation of Phase I trials, etc. Any ethics of clinical trials are justifiable in the aspects of investigator disclosures, valid consent of the volunteers, competency, and the foremost the morality. Nowadays, a number of foreign medicines are registered for marketing in India, but in the legal framework, all drugs are required to register first and data from a human trial with a minimum of 100–140 Indian population are required. In the regulatory framework of the clinical trial, the clinical trial for the new drug/investigational new drug (IND) cannot be started without the permission of DCGI. Clinical trial permission is granted after applying with Form "44", along with the requisite data and fees. Clinical trials are controlled strongly and are under government regulation.

14.3 DRUG COSMETICS RULE, 1945; SCHEDULE Y, AND AMENDED IN 2005 (II), FOR A CLINICAL TRIAL

Schedule Y is the established regulatory framework of clinical trials for imported drugs, sale of new drugs, new chemical entities, and new drugs. Despite some expert opinions that Schedule Y boosted and promote generic drugs and provides equal opportunity, foreign pharmaceutical companies enter the Indian market to face regulations and obligations of strict patent rules. Schedule Y is based on a strong regulatory framework concerning the government control of conducting clinical trials, the process of application, the financial obligation of the investigators and sponsors, compensation of any the boundary of responsibilities of the sponsor and investigators, and the power of the independent IEC holding the morality or injury and power of attorney to safeguard the interests of volunteers. Clinical trials can only be initiated after obtaining written permission from IEC and DCG (I). After approval of the DCG (I) and IEC, followed by entry in clinical trial registry, any clinical trial can be started as per Schedule Y. After the thalidomide disaster, the Indian government is very particular about imported drug, which cannot be sold in India after the conclusion of at least open Phase I trials and obtaining the trial license called a "T" license which within the Schedule Y has to follow. Similarly, the importing of the biological product required a "No Objection Certificate" (NOC). Once the study is approved by DCGI, the T-License and NOC are granted within 2–4 weeks, and multiple shipments can be allowed in a year.

14.3.1 Rule 122 E

This section deals with the new drug and the new drug definitions are quite broad. When a drug is imported and introduced into India, or an established drug is in a new indication of treatment in new dosage forms and fixed-dose combination, these are considered as new drugs.

14.3.2 Rule 122 DAA

The systemic study of new drugs, especially focusing on the methodology for processing new data generated in clinical, pharmacodynamic, and pharmacokinetic studies. These studies are critical in the evaluation of the safety and efficacy of new pharmaceutical products. Also, this section is important for the reporting of adverse reactions, determining the safety and efficacy of the new drugs or verifies the new molecules as under section 122 E.

14.3.3 Rule 122 DD

This section is particular to the registration and functioning of the IEC. To ensure the highest level of volunteer safety and protection, the responsibilities are collectively shared by the sponsor and investigators. The IEC is the safeguard of patients and the laws and rules implemented by the regulatory authority on behalf of the volunteer safety and interest. Also, reasonability is arisen to generate quality data and a valid clinical trial. Approval by DCGI is mandatory for any clinical trial. After obtaining the application in the specified format with the requisite data and fees, both are reviewed by the committee and approval is subsequently granted. Requisite documents and information are needed to be submitted to the Drug Controller General of India (DCGI) through the IEC when seeking approval for a start clinical trial in India. This process is in line with regulatory requirements to ensure that clinical trials are conducted ethically, safely, and in compliance with the law. The special type of patients involved in the trial are required to obtain special permission with proper justifications and the concurrence of IEC and DCGI. The special type of patients or volunteers are pregnant women, nursing women, children, elderly patients, and other special types of people like tribal populations, soldiers, and others with predetermined diseases such as kidney, hepatic, or other organ failure or terminal patients who are in the specific treatment. Under this rule is also described the composition of the IEC, which includes a minimum seven-member committee. The committee is also constituted at least 50% of people who are outside the Institute and the chairman must be from outside of the Institute and other members are a clinician, a statistician, a medical scientist, a social scientist, a legal expert, and a common person from the community. The IEC is also responsible for forwarding the reports of any adverse reactions or deaths that occur to the DCGI and accordingly deliver financial compensation within 21 calendar days of the unpleasant event. Normally an IEC constitutes for three years from the date of issue unless the committee is dissolved.

14.3.4 Rule 122 DAC

Rule 122 DAC pertains to the adherence to regulatory and ethical protocols governing the data supplied for clinical trials, as well as the measures implemented in response to instances of noncompliance. The Central License Approving Authority (CLA) has the authority to conduct inspections on sponsors, which include their employees, subsidiaries, and branches, as well as their agents, contractors, subcontractors, clinical trial locations, BA/BE centers, and investigators. These inspections are carried out to ensure compliance with regulations, and it is expected that sponsors provide sufficient responses to the inspectors regarding the execution of clinical studies. The CLA has the power to refuse or terminate studies and prohibit individuals and entities involved in clinical trials, such as investigators, sponsors, employees, subsidiaries, branches, agents, contractors, and subcontractors, from participating in future trials. It is of utmost significance that every stakeholder comprehends their respective responsibilities and adheres closely to the regulatory obligations, particularly in instances where a comprehensive grasp of good clinical practice (GCP) and training is required for all stakeholders.

14.3.5 Rule 122 DAB

Safety establishment of any new drugs is a major concern. This section describes the adverse drug reactions or any unwanted events during the clinical trial, including the directive of compensation. Accessing any adverse drug reaction could be established through the Phase IV clinical trial (post-marketing surveillance) and this section is very important to access the adverse reactions. Most adverse drug reactions are reported through the IEC at the time of Phase I–III clinical trials, otherwise in Phase IV or anytime the untoward adverse reaction can be directly reported to the DCGI or USFDA. Compensation is evaluated according to the severity of the adverse reaction to the drug. Extreme cases can be considered the death of the patients and other adverse reactions can be considered the hospitalization of the patients in case of the clinical trial are conducting in the out-patient and the prolongation time of the hospitalization is considered in both cases in-patients or out-patients volunteer. Sponsors bear all responsibility for expenses related to the treatment of volunteers. Any disabilities, including permanent or temporary incapacitation, occurring during the clinical trial are to be considered and compensation accordingly framed. Other adverse reactions including prenatal toxicity, congenital abnormality with abnormal birth, birth defects, or other risks associated with life-threatening conditions are also considered at the time to be framed charges for compensation. In addition, the compensation charge and the responder are considered the sponsor or the sponsor's representative by the investigator. Mainly charged framed accordingly the gross violation of the approved protocol, scientific negligence, misconduct, clinical trial procedure, spurious drugs involved clinical trial, false claiming of the therapeutic effects and the clinical trial any placebo used the additional charge can be formed including the use of placebo and improper patient care those are involved in placebo therapy.

Informed consent documents (ICDs) are among the important documents to frame any charges in case of adverse reaction during clinical trial and compensation issues. According to Appendix V of Schedule Y, ICDs clearly direct that if any injury is reported during the clinical trial, the subject should get free treatment as long as required, and in cases of death or disability, full compensation is to be provided. ICDs should mention the right of volunteers and informed volunteers about compensation in case of trial-related injury or death. The addresses of investigators and sponsors are to be mentioned in the ICD.

14.4 ROLE OF THE CENTRAL DRUGS STANDARD CONTROL ORGANIZATION (CDSCO) FOR CLINICAL TRIALS

One of the key responsible regulatory authorities in implementation of clinical trials in the fragrance of legality Ministry of Health and Family Welfare (MoHFW), which is responsible overall authority for maintaining the standards of drugs, cosmetics, diagnostics, and devices in India to ensure their safety, efficacy, and quality by implementing different rules and regulations. The Central Drugs Standard Control Organization (CDSCO) is the regulatory authority under MoHFW to control of the clinical trials standards procedures and standards, and is the final authority to grant licenses for sale of drugs in India. CDSCO also controls the import of new drugs and approved drugs for trial and marketing. Indian GCP guidelines framed by CDSCO in 2001 and CDSCO strongly regulate the implementing of the GCP. Every clinical trial is required to send the adverse drug reaction (ADR) and GCP reports after completion of the year or upon completion of the clinical trial.

CDSCO also takes initiative to strongly monitor ADR throughout the nation. A large number of medications are involved in coordination of particular ADR monitoring centers (AMCs) and private medical colleges and nursing homes are also included. The Indian Pharmacopoeia Commission (IPC), Ghaziabad was set up as a national coordinative institute to provide all technical support. The pharmacovigilance database is prepared and reports are included in the WHO-Uppsala Monitoring Center ADR database.

Another important step was initiated by CDSCO for inspection of the clinical trial sites by USFDA-trained Indian inspectors and other regulatory experts to ensure the quality of the trials. In 2005 amendment to Schedule Y, CDSCO framed the clinical trial process according to international standards. In two aspects, this amendment is very important. First, in this amendment, the definition of the clinical trials Phase I–IV and their time frames changed accordingly in the approval procedures. Second, more clear responsibilities of the sponsors and investigators are well documented. CDSCO also gave instructions, provided guidance, and laid procedures of the clinical trial inspections. All information of the related clinical trials is also mentioned by the sponsor, investigator, and CRO, and all should be notified accordingly of any changes.

14.5 ROLE OF DCGI IN CLINICAL TRIALS

Approval authority of any clinical trial and application must be made through the recommendation of the IEC. Further, the amendment establishes that those running clinical trials are required to notify the DCGI in writing with the concurrence of IEC approval within 30 days of implementation. After approval, no alternations are permitted except stopping the use of hazardous materials and reporting serious health hazards to the participating subjects. If any adverse drug reaction occurs, it must immediately be reported to the Directorate General of Health Services, CDSCO, DCGI, by the sponsor, and all trial participants within 14 calendar days. Also, this adverse reaction is to be reported to all regulatory authorities of any other countries where the drug has been already permitted to trial. The DCGI also categorized the approval processes of foreign drugs or imported drugs. In category A, trial applications for drugs already approved in the United States, Germany, South Africa, Britain, Canada, Australia, Japan, Switzerland, or member countries of the European Medicines Agency (EMA) are considered fast-track applications and approval is given within a short period (2–4 weeks). In category B, if the drug approved to other countries and approval taking a long time after investigations and the times take almost 12 weeks after the application considered under category. Most of the global trials considered as a category A and the never category B shifted to category A drug trial. As amended in 2005 Schedule Y, every sponsor is pro forma required to submit the progress of the drug trial every six months. Any discontinued study should be reported to DCGI within three months with the description of the trial, dose, number of patients and duration of exposure, details of adverse drug reactions, and justification for not pursuing the new drug application or the reason for discontinuing the study.

14.6 TYPES OF CLINICAL TRIALS, ACCORDING TO DRUG APPLICATIONS

14.6.1 Adjuvant Therapy

The trial is initiated for evaluation of a drug's efficacy to stop the recurrence of or prevent the occurrence of a disease after successful treatment with surgery, replacements of organs, radiation, chemotherapy, or any other means. Most of the oncology, trauma, or accidental cases have come in this trial. The endpoint of the trial is that without any clinical signs of disease the survivability or life span after the intervention.

14.6.2 Neo-Adjuvant Therapy

This trial is conclusive of the effective treatment or the efficacy of the treatment before the adjuvant therapy or effective treatment, such as for example pre-ambulatory medication before the final replacement surgery or pre-oncologic treatment before surgical oncology. The endpoint of the clinical trial is to draw conclusions about development and improvement before considering the patient's surgical options.

14.6.3 Surgical Intervention Trials

New surgical techniques or surgical protocols which are scientifically effective are to be demonstrated are fall in the clinical trial. The surgical procedure should be an ideal extent, quick, minimum intervention to the patients and short time are tested through the clinical trials system.

14.6.4 Supportive Care Trials

A study of a designed therapy that seeks to increase the quality of life of the patients, including nutritional interventions and psycho-social options, is known as a supportive clinical trial. These types of trials are more expanded to study of side effects and the treatment of side effects. Diseases including diabetes, chemotherapy, and neurological disorders come in this category. The trials are based on how to achieve the reduction of pain and side effects of the disease and to help the supportive measure to the families or caregivers of patients to overcome the disease.

14.6.5 Prevention Trials

Preventive studies will be offered to people who are at high risk of developing a condition. The medication lowers their risk of developing diseases. Persons who handle communicable diseases like human immunodeficiency virus (HIV) and tuberculosis (TB) patients should be given the preventive therapy in low doses which might reduce their risk of developing the disease.

Walking can sometimes reduce the risk of obesity or cardiovascular disease, and these methods include prevention therapy or lifestyle modifications therapy.

14.6.6 Early Detection Trials

Epidemiological studies such as screening for diseases such as Japanese encephalitis (JE), malaria, HIV, hepatitis, cervix cancer in women, etc., by any intervention are known as early detection trials. This form of clinical trial is undertaken in a normal population of people.

14.7 EXPERIMENTAL DESIGN OF THE CLINICAL TRIAL

The success of each clinical trial is determined by its design. The clinical trial design is determined by how efficiently questions are to be answered, the accuracy of data generation, and the variability of experimental units.

Investigators develop and design a specific dosage type or intervention for clinical trial participants. Interventions are a broad phrase that encompasses a variety of changes such as medical items, medications, or technologies, as well as changes to participants' routines such as food and data-gathering processes. The clinical trial is meant to ensure the scientific integrity of the experiment, as well as its credibility with clear evidence of safety and efficacy. The active medications are compared with a placebo (without active ingredients) or without intervention to see if the new experiment is beneficial or harmful.

14.7.1 Observational Study

This type of clinical study examines the nature of cause-and-effect relationships. Participants are not allocated to any group or with any specific treatment, and the results are concluded by asking questions, some clinical parameters, or some measurement. Most of the sample survey falls under the observational study.

14.7.2 Case Studies

Case studies are defined as a minimum of the patient diseases intervention or treatment in clinics or hospitals, and they are not considered clinical trials. The case study reveals information about the disease or treatment patterns.

14.7.3 Case Reports

A single patient's disease intervention or treatment in clinics or hospitals is known as a case report, and because there is no scope of the control group, it cannot be deemed a clinical trial due to statistical inconsistency. The case study, on the other hand, clearly presents a hypothesis of the disease pattern.

14.7.4 Case-Control Studies

Case-control studies involve chronic exposure to the environment and effects that can be linked. The results are extended retroactively, and reports are reviewed to reach a conclusion, such as linking cigarette exposure to cancer, though this type of study cannot provide proof of concept and cannot be directly associated due to the limited statistical analysis and lack of establishing cause-and-effect relationships.

14.7.5 Cohort Studies

A cohort study is an observational study in which some persons in a cohort (a small group of people with something in common) who have been exposed to a certain causative agent or therapy are compared to a similar group of persons who have not been exposed, such as tobacco smoking among working and nonworking young people. When compared to the randomized clinical trial, this study has the smallest statistical significance.

14.7.6 Cross-Sectional Studies

Cross-sectional studies examine the impact of disease exposure. The impact of COVID-19 pandemic traumatic stress on the population is a classic example. Cross-sectional research is performed to determine the sensitivity and specificity of novel diagnostic tests.

14.7.7 Single-Arm Trials

This is the most basic trial design in which there is a limited scope to access the effects of placebo since placebo treatment may be unethical or limited in scope. The design is straightforward; to understand directly the effects of the intervention on the research population, a baseline is derived at the start, which may be used as a control group. This type of study is critical for validating Panchakarma procedures, Agnikarma, Ksharasutra, and so on.

14.7.8 Crossover Trials

This is a sort of random control experiment in which the control or placebo group and treatment group are assigned at random. After a trial period, the allocated group reversed after certain buffer periods, which could be a different protocol, and compared the results. Individual variances and differences are eliminated with this study.

14.7.9 Factorial Trials

Factorial trials are designed when there are two variables or two trial procedures, and measurement involves knowing the effects of treatment on the subject group. Assume that A and B are two separate treatments, and the factorial trial requires four groups. Group I will be treated with A, Group II with B, Group III with A + B, and Group IV with neither compound.

14.7.10 Non-Inferiority Trials

Also known as active-controlled trials, these trials compare data from existing treatment procedures to determine whether the new treatment is superior or inferior to the current treatment. Using placebo groups after establishing non-inferior trial findings is unethical.

14.7.11 The Parallel Design of Placebo-Controlled Trials

The psychological consequences of any therapeutic technique cannot be denied, and they can occasionally play an essential role in any medication study. Placebo is a drug/therapy/formulation imitation that has no pharmacological action but instead has a non-specific psychological mechanism. Thus, a placebo-controlled design in a double-blind trial is crucial to grasp any specific effects of drugs/therapy rather than any psychological effects—but understanding the right results requires large sample sizes.

14.7.12 Three-Arm Trials: Placebo and Active Control

This type of trial is comparable to the parallel design of placebo-controlled trials, except that an extra treatment is or active groups are included, results are compared with the placebo group, and all trial participants give good internal evidence of sensitivity. This type of experiment is conducted in a double-blind style with a high number of samples.

14.7.13 Add-On Study

Data from this type of study have a high level of significance, making it comparable to the additional treatment group. If no clear treatment is available and any other treatment is viable, this type of investigation is conclusive. The majority of repurposed medications are employed in this experiment, with COVID-19 therapy being a famous example. If there are no adverse reactions or drug interactions, alternative treatment is usually used in addition to the normal treatment to improve outcomes.

14.7.14 Replacement Study

This type of study is created randomly and aims to gradually withdraw from standard treatment by replacing it with placebo. The results of such a trial provide effective data on the baseline status after discontinuing conventional treatment.

14.7.15 Early Escape Rescue Treatment

Many medications and therapies are required to assess the risk-to-benefit ratio. Some medications are exceedingly strong and have a low therapeutic index; such drugs can occasionally aggravate disease and should be discontinued. This sort of trial assesses the tested medication's short-term acute exposure.

14.7.16 Limited Placebo Trial

A placebo-limited trial is a time-limited trial using placebo for the development of effective data at low exposure. The placebo group is gradually replaced by active medications or active groups.

14.7.17 Additional Doses Design

The dose response of two separate active medicines is determined by randomly assigning participants to parallel groups under a similar treatment procedure. This design generates useful data.

14.7.18 Randomized Withdrawal

The purpose of this experiment is to determine the status of disease decrease or exacerbation, as well as long-term efficacy. To understand the effects of the treatment, participants were randomly assigned to treatment and placebo groups for a set period of time and then partially withdrew randomized in a similar technique.

14.7.19 No-Treatment Concurrent Control

In a double-blind trial, the no-treatment and treatment groups are available, and the results of the comparative research are concluded.

14.7.20 External Control (Including Historical Control)

A multicentric trial design is used to conduct this type of experiment in at least two sites. It is a compression study, with trials taking place concurrently. External control refers to one location trial, while internal location refers to another.

14.7.21 Control Groups in Clinical Trials

Control groups can be described as based on the nature of the trial, such as when a control group receives standard drug in comparison to a clinical trial or when the efficacy of the drug is to be determined. However, the terminal disease placebo clinical trial cannot be considered. The evaluator can quickly access the impacts of pharmaceuticals, adverse effects, or drug effectiveness, and investigators can accordingly improve further. Furthermore, the effect of medications can be examined for a specific gender or group of people, or adverse effects can be exaggerated for a specific gender or group of people.

Another significant function of the control group is to counteract the biasing effects of experiments that only have one control group. Keeping the control group is critical in an epidemiological clinical trial or illness assessment.

14.8 RANDOMIZED TRIALS

Avoiding bias in the reporting of the treatment performed in the unit or block by randomly selected volunteers. The final results collected from the volunteers are the sum of all outcomes and are not dependent on the individual results. In addition, all external influences are spread equally in this process, with no influence on individuals. When randomization is performed, the factors influencing the results must be discovered. Therapeutic interventions must be determined without regard to unknowing bias, and randomized approaches must be used.

Randomized controlled trials (RCTs) evolved from the randomized trial to more scientifically adopt the set technique and are typically used for the evaluation of drugs and medical procedures. Random patients are allowed to be exposed to the treatment in the RCT, and the impacts in terms of results are to be calculated in accordance with statistical equivalency. RCT is regarded as one of the greatest scientific approaches for determining outcomes after eliminating all sources of bias and calculating average efficacy while also accounting for negative effects. The following are the implications of the RCT on data reliability.

14.8.1 The Human System or Any Biological System's Response is a Complicated System

1. Responses to any stimulation are determined by various receptors, environmental circumstances, and other factors. Response to a biological system cannot take into account important outcomes and scientifically unacceptable evidence. As a result, studies should be conducted on a larger number of subjects in order to get more acceptable scientific data.

2. Sometimes the treatment's effects are too weak to define, and in order to achieve more acceptable results, the small effects amplify in a larger population group.
3. In order to achieve better outcomes, the RCT must complete the treatment criteria.
4. The RCT is to be carried out in order to overcome the patient's psychological consequences. In an RCT, half of the patients are chosen for therapy, and this group is known as the treatment group or investigational group, while the other groups are known as the control group. A comparison is done between the treatment and control groups.

14.8.2 Procedure of Randomization Method

1. Simple random: This is a form of allocation in which each sample has an equal probability and is a subset of the population that provides external validity. The random sampling is generated using an online or Excel spreadsheet.
2. Random allocation: This is the process of randomly assigning distinct groups to an identified sample. A random allocation procedure is one that is assigned with equal probability.
3. Randomization block assignment: This is a method effective for assessing the effects of medications in the interim. The block assignment is completed in two or more groups, and the assignment is completed with the same sample size. The blocks are randomized and created, and the same units are employed as a control, as well as an active or treatment group.

14.9 OPEN TRIALS

An open trial occurs when the investigators determine the type of treatment, the details of the treatment, and the patients to whom it is applied. In most instances, open trials have minimal value in clinical studies or pharmaceutical evaluations because the study is considered biased. This type of clinical trial cannot measure placebo effects.

This trial is unavoidable in some circumstances, such as surgical techniques, drug bioequivalence studies, and so on, whereby concealment is unethical and full information of the experiment must be provided to patients and investigators.

14.10 BLIND TRIALS

This is a method for overcoming the biasing effects of individual factors at the time of the experiment. When comparison studies are conducted, the results may be altered if the type of medication or treatment—and who received it—are disclosed. In this scenario, both the investigator and the patients have a psychological inclination to skew the results for that experiment, which is maintained "blind" to the patients and investigators.

14.10.1 Single-Blind Trials

The blind study is now used to overcome the biasing of results. A single-blind study is when the nature of the prescribed medicine is not disclosed to the subjects. Much research has found that when investigators are aware of the treatment regimen and inadvertently indicate the therapy or particular care to the patient, the results are influenced or biased.

14.10.2 Double-Blind Trials

Overcoming the investigator's bias in a single-blind trial is frequently referred to as a double-blind clinical study. This is the most widely accepted clinical trial, during which the investigators are not concerned with treatment details and the possibility of disclosing information to patients is minimal. Furthermore, this form of study is highly rational in terms of the frequency or distribution of the patients in terms of age and gender.

14.10.3 Triple-Blind Trials

The subject, researcher, and statistician are all blinded in this design. In this instance, everyone is blind, and the probable results for each group cannot be predicted.

14.11 UNBLINDED/NON-RANDOMIZED CLINICAL STUDIES

Unblended or non-randomized clinical trials are those in which patients are assigned to the control and treatment groups at random.

14.11.1 Crossover Trial Designs

A crossover clinical trial occurs when no additional groups are assigned to the trial and the placebo and therapy are administered to the same patient. Typically, a buffer time is provided between the placebo and the actual treatment.

14.11.2 Factorial Trial Designs

Factorial trials are those in which two or more therapies are assessed in groups of participants in a single experiment.

14.11.3 Hybrid Trial Designs

These trials are often a combination of historical and randomized control studies. For control data, a limited number of volunteers are kept, whereas a significant number of patients are kept for the test drug.

14.12 THE RULES FOR BREAKING OF THE CODE FOR RANDOMIZATION CLINICAL TRIALS

14.12.1 In an Emergency Situation

Provisions for research un-blinding are allowed in emergency circumstances to learn the root cause of any adversity. In an emergency, the investigator or a representative of the investigator must break the code using a well-defined mechanism and access the code in a sealed envelope within 24 hours. Notification is necessary within 24 hours of the sponsor breaking the code, and the investigator may not break the code until authorization is gained from the sponsor.

14.12.2 End of the Study of the Trial for Breaking the Codes/Un-Blinding Code

The blinding study code is broken when the trial is completed or with the last follow-up to the participants. The sponsor submits a formal request to the institute head, and it is also essential to verify that the data are entered such that there is no opportunity for alteration.

Typically, such databases or software are compliant with 21 CFR (Title 21 of the Code of Federal Regulations) and validation is performed on a regular basis. The statistician also advocates breaking the code after only comprehensive analysis. Ideally, information about data storage and any code breaking in the event of an emergency should be saved for future reference.

14.13 CONCEALMENT OF VOLUNTEER ALLOCATION

This is a technique whereby neither participants nor the investigator could generate any idea of the trial or which medicine is administered to participants. The placebo mimics the investigation drugs in a similar fashion like dosage form, weight, color, and smell.

The double-blind study is different and more restrictive when the random sequence of codes is generated and allocated to the trial participants. The system-generated codes are sealed in an envelope with aluminum foil to prevent access against the light. The participant is given the code and the corresponding medicine or placebo is administered. The envelopes are kept in the pharmacy for future reference and the SOP to be followed by the pharmacy is to mention the code, dispense date and return date, and used packets or vials for auditing purposes.

In single-blind studies, the investigator could understand the trial and which drugs are received by the participants. In this type of trial, the code is maintained for placebo or treatment drugs. The drugs or placebo codes are encoded on the consecutively numbered sealed envelopes. The codes generated are kept by the sponsor/central funding agency, un-blinding information is disclosed at the time of the emergency, and the participants are automatically revealed.

14.13.1 Standard Operating Procedures for Decoding

SOPS are established for the decoding procedure. The justification of the code breaking and the identity of the person for codebreaking is to be mentioned, as is the authorization of the person for codebreaking and the bottle or box number of trial medicine.

14.14 ROLE OF PLACEBOS IN A CLINICAL TRIAL

Placebo is a dummy drug with the same looks or format—like size, shape, color, etc.—to be delivered to patients without any pharmacologically active ingredient, but it is an ethical concern when a patient is seeking treatment and plying this belief. To understand the pharmacodynamic effects of trial drugs and the influence of psychology and other extraneous factors, the placebo trial is done, which helps to draw conclusions about the efficacy of the drugs.

Another important issue is the bias factor of conclusions about the results, and the bias of drawing results is a physiological error whereby results are affected by strong likes and dislikes.

14.15 FACTORIAL DESIGN IN A CLINICAL TRIAL

When one or more variables to be evaluated—including the main effect, interactions, or more factors at the time of experiment—this is known as factorial design of the clinical trial.

14.15.1 Main Effects

In the factorial design, the main effects are determined according to the dependent variable. Depended on the variable is considered as exercise, nutritional therapy along medicine. Individual effects are factors, and the cumulative effects of all factors are determined. Here, the prescribed medicine can be considered as the main effect.

14.15.2 Interactions

In the factorial design, interaction is determined from the drugs or drugs with other drugs or foods or hobbies. The drug's effects are varied from the main effect if the cumulative effects are different.

14.16 PATIENT SAFETY IN A CLINICAL TRIAL

Conducting any clinical trial without considering patient safety is a crime and can be prosecuted in the legal framework of India. All clinical trials should follow the laid rules or norms. The Honorable Supreme Court in September 2013 directed that no clinical trial is permitted without monitoring and without the proper mechanism. Every clinical trial should be approved by an IEC and the DCGI. IEC members should be provided with adequate knowledge about the rules, ethics, and safety justified for the trial. Each trial should mention risk and beneficial effects, and it is required that volunteers can withdraw from the trial without giving any reason. In IEC reviews, it is considered that each trial should be ethically correct and the protections of the fundamental Article 21 of the Indian Constitution should be provided. Any untoward incidence or any adverse effects reported immediately should be reported to DCGI and compensation should be fixed according to the type of the injury or disabilities. The IEC has the power to stop the trial any time if the accident or adverse reaction occurs. The SOPs of the experiments, patients' data, consent, and the recommendation should be kept accordingly. CDSCO recently ruled that all documents of clinical trial are auditable.

14.17 TYPES OF CLINICAL TRIALS

Clinical trials are the process for stepwise evaluation to ensure the safety and effectiveness of a new drug or medical device for placement in humans. The phases of the clinical trial each have their own significance and designed accordingly.

14.17.1 Phase I Trials

After the success of pre-clinical toxicity evaluation, the Phase I trial is initiated after according approval of the institutional human ethical committee and compliance with other regulatory obligations including the consent, investigator declaration,

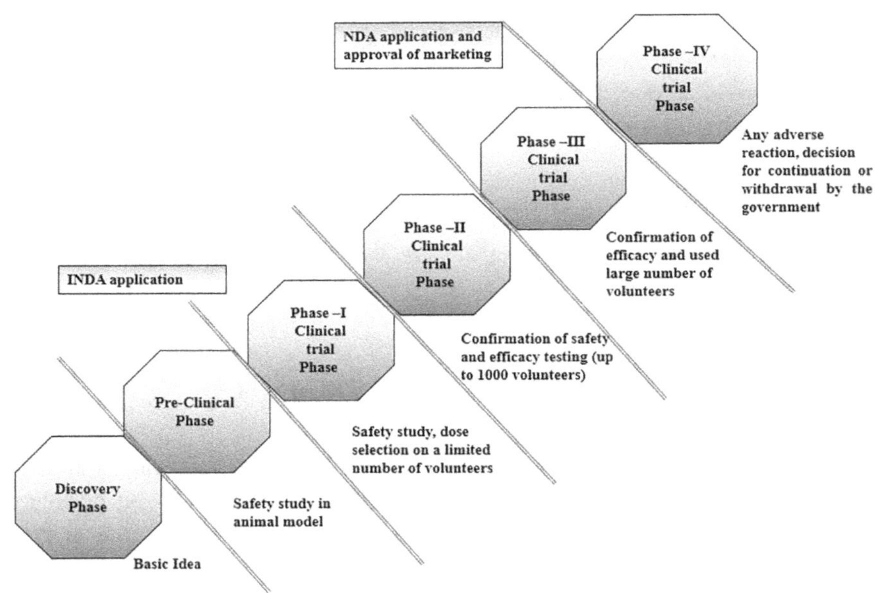

FIGURE 14.1 The process of new drug approval.

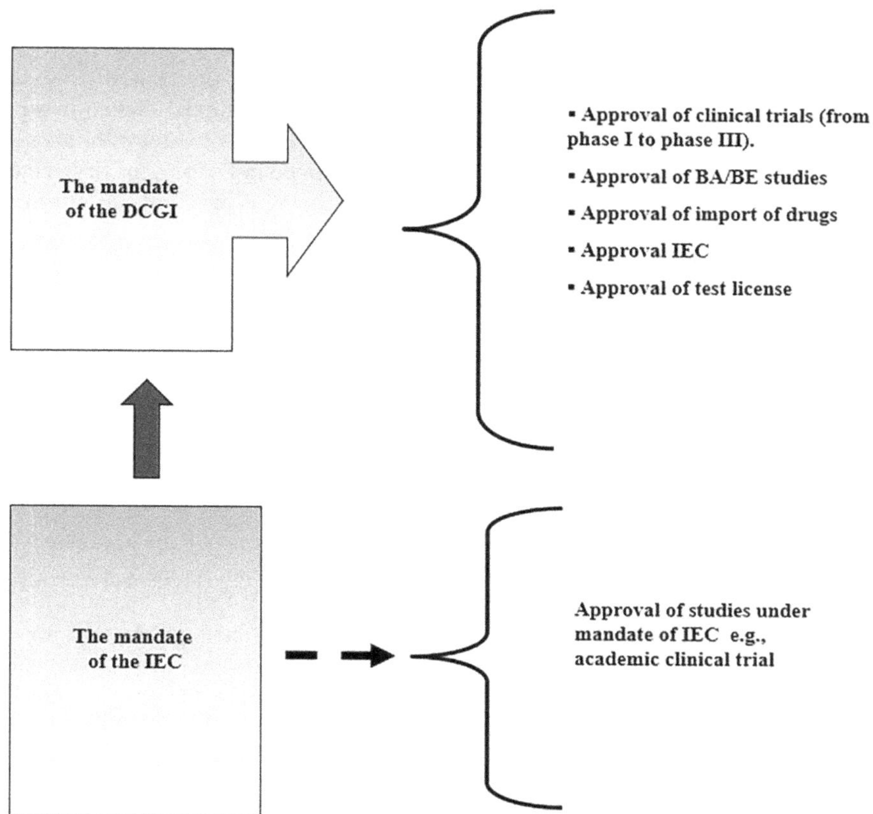

FIGURE 14.2 Relationship between clinical trials, DCGI, and IEC.

and investigator brochure with the clinical trial registry, as per the government norms. Phase I trials provide clues that the new drug, device, or treatment is safe at the particular dose. A small group of people (20–80) is used in this trial, and this phase considers the first step to using experimentation. This stage is confirmatory of dose range study in the safe zone or determination of therapeutic index after data is generated in dose fixation studies of large animals. Also in this phase, the route of administration is determined as to whether the drug has therapeutic or prophylactic value via oral, inhalation, intravenous, or other routes. The trials are conducted in an inpatient clinic or designated hospital. It is very important that the IEC should be approved by DCGI. In Phase I trials, the patients are observed very closely until the half-life of the drug passes for any adverse reaction, and full-time medical staff is required. Phase I study is a broad area, and the establishment of bioavailability of new drugs and bioequivalence studies are also considered in the Phase I trial.

Phase I establishes the safety of the drug and the volunteers are monitored very closely. Only single drugs are allowed to be administered to volunteers in study of a particular disease or illness. The drug's highest tolerated dose, effective dose, and interaction with foods is also determined in this phase. Most of the ADRs are reported in this phase. Pharmacokinetics and bioavailability of the new drugs are also determined in this phase after testing blood at various times. The antiviral or antiretroviral therapy in Phase I is limited in the range of 1–3 weeks for minimization of the risk of resistance. After satisfactory results obtained in Phase I, the drug enters Phase II and the same volunteers may continue after being given some buffer time.

14.17.1.1 Objectives of Phase I Clinical Trials

1. Establishing the safety and toxicity of new molecules.
2. Study of the pharmacokinetics and biovailability of the new drugs, and determination of clues to the therapeutic range and dose adjustment in Phase II trial.
3. To evaluate the drug safety by using small numbers of volunteers, with short durations of exposure and close monitoring of any adverse drug reactions.
4. Establishing the initial efficacy, safety, and maximum tolerance level of the drug.
5. Establishing the metabolic and excretory pathways or active metabolites.
6. Establishing the dose-response variability between individuals, route of administration, and bioavailability.
7. Indication of therapeutic effects, side effects, and concluding the contraindications.
8. To identify the possibilities of the factors that influence the drug actions, including age, gender, genetic differences, and food.
9. To measure or quantify dose/concentration–response relationship.
10. To establish any drug interactions.
11. To establish absorption, distribution, metabolism, and excretion of the drug.
12. To establish the bioavailability or bioequivalence of the drug.

14.17.1.2 Types of Phase I Trials

14.17.1.2.1 First-in-Man (FIM) Trials

This study initiated with a single dose to assess the tolerability of drugs, safety, plasma kinetics, and the pharmacodynamic effects, and to conclude the results by extrapolating with pre-clinical studies. In most cases, the FIM study is an important aspect to overcome the risk associated with the clinical trial.

14.17.1.2.2 Ascending Dose Trials

Phase I trials are initiated with single or multiple ascending doses.

14.17.1.2.2.1 Single Ascending Dose (SAD) Trials This type of dose assignment is very important for a very new molecule and concludes the predictive safe pharmacokinetics or the concentration of the drug in plasma at a safe level. A single dose of drugs is administered to a small volume of patients and observation periods keep long to predict any observed toxic side effects. Pharmacokinetics data are established in a single dose and concluded in the safe region of pharmacokinetics profile. Another group of volunteers is administered escalated doses and the resulting pharmacokinetics data are generated. This process is continued until pre-calculated maximum tolerated pharmacokinetic safety levels are reached or at the point when intolerable side effects start showing up. The dose is known as maximum tolerated dose (MTD) and this is important for adverse drug reaction and pharmacovigilance monitoring at Phase IV clinical trial. MTD determination the dose adjusted 1/6th to 1/8th of the effective dose which predicted at the time therapeutic dose determination. Minimum one week interval or buffer is given to volunteer to prevent the cumulative effect in the Phase I trial.

14.17.1.2.2.2 Multiple Ascending Dose (MAD) Trials Pharmacokinetics and pharmacodynamics are the most important aspects in Phase I clinical trials. When establishing the new drug and how it acts in the body or better understand the pharmacokinetics and pharmacodynamic clinical trial, multiple doses of the drug are useful. In this study, the escalated manner in which the drug is administered, and at the time when the body fluids—including blood—are withdrawn and analyzed, gives the clear picture of the drug's interaction with the human body. The effects of food on drug absorption in Phase I trial is also considered. The effective of pharmacokinetics in the presence of food and an empty stomach is also determined and the drug is prescribed accordingly. All of the parameters are determined in Phase I trial.

14.17.1.4 Important Documents of Phase I Trials

The clinical procedure and investigator brochure is an important document in the Phase I trial. The patient's declaration, consent, purpose of the clinical trial, risk and beneficiary disclosures, and insurance of the patient are documented.

14.17.1.5 Safety Data of Patient

Before any clinical start to understanding the nature of the molecules from published literature or relevant medical data and drawing conclusions about the safety aspects, the medical records of the volunteer are to be determined before entering into a clinical trial. The first dose administered to close of the observations until several half-lives pass in the treatment.

Safety aspects of the biological are more different safety concern and after TGN1412 tragedy the pharmacovigilance are becomes stronger. Monoclonal antibodies and other proteins act on off-target or many targets, such as the complement system or Fc receptors, and other non-specific targets like antigens. Safety monitoring must record very closely because of the risk of anaphylactic or infusion reactions; delayed hypersensitivity reactions are sometimes also reported.

14.17.2 Phase II Trials

Larger volumes of volunteers are deployed in this type of trial, with 100–1000 volunteers typically involved in such studies. This trial evaluates the drug efficacy and safety, and establishes its mechanism or mode of action, though this trial is considered a safe study as compared to Phase I trial. Design of a Phase II trial is quite technical and it is observed that no "biasing" reports should be made. Phase II trials are conducted in the following manners.

1. Comparison between subjects: The results are compared between the treatment group and standard drug treatment group. In this study, randomized volunteers are selected.
2. Subject comparison within the group: Nullifying the patient factors here, the drug is tested in the same population groups with drugs and placebo groups alternatively.

Sometimes in Phase II trials, placebo trials are very common even though controversies exist regarding placebo therapy in the concern of drug efficacy, particularly when the patient is in disease conditions. Determination of the therapeutic index, dose determination, and associated adverse reactions in the Phase II study is very important. Case study is designed for the evaluation of the safety and efficacy for a particular group and the phase studies designed for the safety studies known as Phase IIA, while study involved dose determination called Phase IIB.

14.17.2.1 Objectives

1. Confirmation of Phase I study of the safety of drug evaluation, therapeutic index, and any adverse reactions.
2. Dose determination and showing maximum therapeutic efficacy.
3. The efficacy of a drug in different age groups requires a comprehensive approach that takes into account physiological differences, disease characteristics, pharmacokinetics, pharmacodynamics, dosing adjustments, clinical trial design, comorbidities, and patient preferences.

14.17.3 Phase III Trials

Efficacy and adverse reactions to a drug undergoing trial are amplified in a large volume of patients numbering 1,000–3,000. Phase III studies are known as the last hurdle for the authorization for marketing. Phase III study also requires understanding of any alternative or better treatment in comparison with a current standard or established treatment. Phase III trials are considered the most complicated, expensive, and time-consuming, and these drug trials take place in patients with chronic cases in a multicentric place. Phase III trial is considered as a multicentric trial for better evaluation of results. Regulatory approval of Phase III clinical data is very important, along with the other data which are required, including the pre-clinical toxicities data, manufacturing process, validation formulation details, and shelf life for regulatory approval. Sometimes the Phase III trial is extended to a Phase IIIB trial to ensure obtaining necessary additional safety and efficacy data. Successful Phase III data are sufficient to get approval from the regulatory approval.

14.17.3.1 Objectives

1. Additional data for drug safety and efficacy in a large volume of patients.
2. Conclusive treatment and efficacy of therapy after comparison to patients treated with standard drugs.
3. Evaluation in pediatric patients with new indications of dosage forms or new formulations.
4. Evaluation of the drug in the special populations and generation of clinical data.

14.17.4 Phase IV Trials

Phase IV is known as post-marketing surveillance, and the regulatory agencies conduct this type of trial or to get reports of any adverse drug reactions after marketing the drug following Phase III trial. The market drug was sometimes withdrawn within a short time if it were eventually reported that the drug demonstrated side effects or showed any adverse reactions. Phase IV studies have no limits on numbers of volunteers or time, with drugs trialed for a long time and tested in hundreds to thousands of people in multicentric places or countries. Phase IV trials are very useful to indicate the mode of action, drug interactions with other drugs, and the drug's effectiveness in categorized patients or certain populations.

14.17.4.1 Sample Size

The basis of Phase IV studies is to get information about the importance of clinically studied groups. A sufficient number of volunteers are recruited for acquiring the data and achieving validation. The sample size or participation numbers are calculated according to the outcome or target for study, null and alternate hypotheses, type I and type II error, or test statistics. The sample size calculation also considers many assumptions like mean or variance of the continuous outcome of each study arm, dropout rates, withdrawal or crossover to the other study arms, any protocol violations or missing data, applied statistical methods, etc.

14.18 IMPORTANT ISSUES FOR CLINICAL TRIALS

14.18.1 Participant Inclusion Criteria

Inclusion criteria set the parameters for subjects and determine eligible candidates who can participate in clinical trials which may be based on age, gender, or other clinical criteria. Participation is to be taken care of so that all participants shall provide consent to participate in the trial.

14.18.2 Participant Exclusion Criteria

Criteria for the ruling out of participants for inclusion in the trial should not be on the basis of the failure of prerequisite conditions. The exclusion criteria may be medical condition or laboratory or physical finding which does not comply with the study parameters. Such conditions could be, for example, any hypersensitive reaction, history of addiction to drugs or alcohol, compromised immunity, pregnancy or lactation, or any other non-compliance which may affect the study.

14.18.3 Withdrawal Criteria

The right of participants reserve to withdraw from the study without showing any reason and any time is as per their wish. However, any adverse reaction, clinical abnormalities, any non-favorable medical condition or situation, or illness that is not favorable to the participant may lead to withdrawal from the study. The investigator shall note any adverse reaction or concurrent illness or any medical abnormality during the trial, and this shall be reported to the adverse drug reaction authority. It is also the duty of the sponsor to take all measures to stabilize the participants from any adverse reaction or concurrent illness during the trial.

14.19 PREREQUISITE STEPS OF CLINICAL TRIALS

14.19.1 Screening

Screening is one of the techniques for identifying participants in the trial according to their eligibility based on inclusion criteria of the study. It is mandatory to select the criteria of the participants and include the name in the register.

14.19.2 Visits during the Trial

The chronological enrollment of the participants is to maintain in the register for any clinical study. Before initiation of the study to enrolled participants is mandatory to visit by the investigator to coordinate the study in order is known as baseline visit.

Successive visits are to be made according to the prescribed time to ensure study protocols are in order, known as follow-up visits. The visits are followed at the end of the study by final study visits to ensure the trial has been completed as per protocols. There are also provisions for rescheduled visits and early termination visits. Rescheduled visits are applicable when the participants or the authority of the institute are not available and the date must be rescheduled to maintain the trial. Any reported adverse reaction related to the study and any non-compliance of the study may cause it to be terminated. The sponsor and IEC must be informed within 48 hours of termination of the study.

14.19.3 Clinical Evaluations

All clinical evaluations are done in laboratories accredited by the National Accreditation Board for Testing and Calibration Laboratories (NABL), with a similar test considering establishment of the efficacy of the clinical trial in a similar fashion to all clinical trial centers. Thus, uniformity of the test quality is preferred by using similar NABL-accredited laboratory reagents, procedures, and chemicals. The important parameter is also to establish the safety of the drug molecules were the periodically testing is required safety of the drugs.

14.19.4 Statistical Analysis

Statistical methods are most important to understand the effects of treatment, confidence intervals, or levels. The block design, volunteer recruitment, and test methodology are adopted to conclude the effects of the treatment or clinical data. It is also necessary to mention to fail or missing the treatment or more than predicted deviation, etc.

14.20 CLINICAL TRIAL OF AYURVEDIC THERAPIES

Ayurvedic preparation is different than modern medicine because its treatments are composed of many ingredients. Thus, a single parametric of the chemical trial could not expect in the ayurvedic clinical trial. Ayurvedic medicine and modern medicine indeed have different approaches to treatment, and this includes differences in the composition of treatments. Ayurvedic preparations often comprise multiple ingredients, including herbs, minerals, and other natural substances, and are prepared based on Ayurvedic principles.

In Ayurvedic clinical trials, it can be challenging to follow the same parametric or design as modern clinical trials due to these differences in composition and treatment philosophy. Ayurvedic clinical trials often adopt designs that are more suitable for evaluating the unique aspects of Ayurvedic treatments.

14.20.1 Systematic Reviews

Most of the herbal or plant medicine claimed in many ways while the Ayurvedic drugs or compound formulation has a limited publication. Systemic reviews are based on the published literature and the credentials of the literature are evaluated against certain pre-fixed criteria by the experts like sound methodology and published literature.

14.20.2 Meta-Analysis

Meta-analyses are a type of systematic review for accessing correlation with confidence rank of the positive interventional studies by the valid studies. This study undertakes the compilation and examination of various findings pertaining to a specific subject matter. The outcomes of these investigations are then combined, and illustrative examples are provided. This study aims to conduct a meta-analysis of randomized controlled trials (RCTs) like example to evaluate the anxiolytic impact of Ashwagandha. Empirical investigations focusing on Ayurvedic medications have the potential to provide essential substantiation for endorsing the utilization of Ayurvedic therapies inside the scientific community. The level of rank in order risk correlated with the textual evidence and compared with the hierarchical model with a long history of clinical experience with the Ayurveda medicine.

14.20.3 Reverse Pharmacology Design

This concept reveals scientific facts rather than investigating the mechanism of action and effects of Ayurvedic medicine to humans which is claimed through the literature. It is the reverse concept of the drug discovery, whereby the discovery follows clinical effects to establish findings in the laboratory.

Phase I reverse pharmacology studies are limited to only collecting the clinical data of standardized Ayurvedic medicine, while Phase II of reverse pharmacology data is limited to target-specific activity of Ayurvedic medicine and extrapolates the tolerable safety dose range and any drug interactions. Phase III of reverse clinical pharmacology activities are rationalized with modern science to establish the effective dose, toxicity to target organ (if any), efficacy, and safe dose.

14.20.4 Black Box Design

This is a particular model applied to Ayurvedic medicine. Ayurvedic medicine is not based on a deep scientific approach; rather, Ayurvedic drugs are based on the practical and theoretical knowledge of traditional medicine which is generated from ancient practice and literature. Ayurvedic medicine is not a single molecule, but rather is a combined molecule and it is believed that that components of the individual molecules have synergistic properties. Thus, clinical trials are based on the comparison of the two groups like placebo and standard treatment or Ayurvedic medicine.

BIBLIOGRAPHY

1. General Guidelines for Clinical Evaluation of Ayurvedic Intervention Central Council for Research in Ayurvedic Sciences, Ministry of AYUSH, Government of India, New Delhi. (Accessed: 20 March, 2021). Available from: www.ayush.gov.in/docs/clinical_evaluation.pdf
2. G.S.R.227 (E): New Drugs and Clinical Trials Rules, 2018. Department of Health and Family Welfare. Ministry of Health and Family Welfare. 19 March 2019. 1–264. (Accessed: 20 March, 2021). Available from: https://cdsco.gov.in/opencms/export/sites/CDSCO_WEB/Pdf-documents/NewDrugs_CTRules_2019.pdf
3. The Gazette of India, New Delhi: Ministry of Health and Family Welfare. Notification 30 January 2013. G.S.R. 53E. (Accessed: 20 March, 2021). Available from: www.cdsco.nic.in/writereaddata/GSR%2053(E).pdf
4. Saxena P, Saxena R. Clinical trials: Changing regulations in India. *Indian J Community Med*. 2014; 39: 197–202.
5. Bhave A, Menon S. Regulatory environment for clinical research: Recent past and expected future. *Perspect Clin Res*. 2017; 8: 11–16.
6. Mallath MK, Chawla T. Investigators' viewpoint of clinical trials in India: Past, present and future. *Perspect Clin Res*. 2017; 8: 31–36.
7. Davis S, Sule P, Bughediwala M, Pandya V, Sinha S. Ethics committees and the changed clinical research environment in India in 2016: A perspective!. *Perspect Clin Res*. 2017; 8: 17–21.
8. Kadam R, Borde S, Madas S, Nagarkar A, Salvi S, Limaye S. Opinions and perceptions regarding the impact of new regulatory guidelines: A survey in Indian Clinical Trial Investigators. *Perspect Clin Res*. 2016; 7: 81–87.

15 Stability Testing of Drugs

15.1 INTRODUCTION

Many herbal medicines have been used traditionally in India for a long time, and knowledge of herbal medicine has been recorded for a long time. As was already said in this book, there has not been a full evaluation of active phytochemicals because they are complicated admixtures of many chemicals. This is why herbal medicines have not been studied much for their chemical content. Most herbal medicines are made by extracting, distilling, purifying, or fermenting whole plants, parts of plants, or liquids that come from plants. Most herbal medicines break down because of heat, light, humidity, oxidation, hydrolysis, microbial attack, or a combination of all of these. Most herbal formulas were made up of several phytochemicals and degraded by phytochemical reactions, but only first-order reactions. Stability is the determined parameter of a drug substance or drug product to stay the same over time, according to its specifications, so that it keeps its uniqueness, strength, quality, and purity without any degradation. Bioactive content is a key part of monitoring stability, which includes the quality, effectiveness, and shelf life of herbal medicines.

The effects of the environment on a drug substance or product or its active ingredient cause it to lose or change its qualities over time, and a stability study can determine how long the finished medicines will last without degradation. Stability studies also give the active period for the drug substance for the highest bioactivity or product, and the recommended storage conditions for it, so the stability study is necessary for judging the quality of the product.

Real-time stability studies take a long time, and it is hard for the manufacturer to keep track of how long it takes for the drug to naturally lose 90% of its labeled amount at room temperature and humidity. Stability testing of herbal formulations at accelerated temperatures and humidity levels is common, and the results of these tests can be used to come up with a reliable shelf life or expiration date by using certain assumptions or criteria. Modern analytical techniques include spectrophotomography, high-performance liquid chromatology (HPLC), and high-performance thin-layer liquid chromatology (HPTLC), and by following the right guidelines, it is possible to make accurate predictions about the stability of herbal products and their shelf life. Accelerated shelf life for any herbal drug can be predicted in a short time, but it takes a long time to get real stability data. Stability data is also used as part of quality control to make sure that the chemical stability, container stability, and active ingredient stability of the final product are all stable while it is being stored.

Accelerated stability studies are usually done on finished herbal products in simulated or higher temperature and humidity. The results can be used to predict the shelf life or expiration date at room temperature by using certain assumptions and criteria. Putting the finished products through accelerated stress can help find out how the active compounds behave under certain stress conditions and find degradation

DOI: 10.1201/9781003398431-15

products. It is true that every product has a clear life cycle and shelf life, which is when the drugs are most effective.

ICH guidelines for stability studies are most acceptable to many countries, including India, for registration of pharmaceuticals for human use. Stability tests should be done on at least three batches of the herbal products in their natural environments. As part of the regulations, a drug must go through a stability study in testing.

15.2 HERBAL DRUG STABILITY TESTING

Herbal drugs differ in certain ways from active pharmaceutical ingredients or allopathic medicine in terms of stability. As requirements of allopathic medicine or unit dosage forms, uniform mass, coating agents, and color are not required, but chromatographs, fingerprint chromatograms of HPTLC, and moisture content are essential to mention. Most herbal formulations have an assay limit of 5% of the declared amount, and this limit is a limiting factor for standardized extracts. The limiting factors of quantifications in batch-to-batch variations are within 5% and exceeding the limit is required to justify a range of up to 10% or even higher. Sometimes the active or marker compound phytochemicals are in low concentrations and are difficult to detect within the limit. Agro-climatic factors are also significant factors for the marker content and must be taken into account. The multiple ingredient-based formulations and low analyte concentrations are quite complicated. Considering the active ingredient content in the linear range and the method's linearity, the marker's expected content can be tested over a range of 40–160%. The active ingredients are added later, and the proposed reasons are as follows.

1. The evaporation of water or other volatile ingredients causes vehicle loss, whereas hydrolysis or oxidation induces drug loss.
2. Because of chemical reactions, solid dosage form reduces bioavailability for aging and appearance color changes, and it may form toxic and irritant products. Furthermore, liquid dosage forms become unstable due to caking of a suspension or creaming of an emulsion, resulting in active ingredient loss.
3. Prolonged storage causes microbiological damage, which causes instability in both manufactured and fresh products. Storage conditions and shelf life should be specified in any product to ensure efficient inventory control and patient safety.

15.3 HERBAL DRUGS AND INSTABILITY

Stability makes sure that a product keeps its identity, strength, quality, and purity for a certain amount of time. On the other hand, it is assumed that storing or transporting the drug at the given time will not affect the quality. Temperature and humidity are considered to be environmental factors in the instability of drugs. Light and air (oxygen, carbon dioxide, and water vapors) greatly affect the stability of herbal drugs. The stability of herbal drugs is also affected by things like pH, particle size, container, lid, and the presence of other chemicals and solvents. Herbal drugs are unstable in humid conditions, as moisture changes the stability of herbal drugs.

15.3.1 Physical Instability

Herbal drugs are made unstable by the presence of microorganisms, impurities in the container and closure, the phytochemical degradation, and formation secondary metabolites. After being stored in a container for a long time, the volatile active ingredients in herbal medicine can easily evaporate and thereby lose their effectiveness.

15.3.2 Environmental Conditions

The quality and stability of herbal drugs are affected by factors like rainfall, altitude, temperature, their source and how the herbs are collected, storage conditions, and how they are processed, such as by storage, drying, cleaning, and extracting.

15.3.3 Chemical Instability

Chemical instability is the most important thing to know about the stability of phytoformulations. Chemical reactions are imitated by self-oxidation, hydrolysis, and the effects of temperature and/or humidity. Herbal products form crystals, break down emulsion, and are altered through enzymatic and chemical reactions. The degradation process is accelerated as temperature and humidity rise. Herbal drugs break down through hydrolysis by absorbing the moisture at high humid condition. Light can also sometimes make the phytoformulations more unstable by releasing free radicals though light-accelerated reactions.

15.3.4 Complex Mixtures, Variability, and Inconsistency

Most herbal formulations are made up of a number of different parts of plants that are mixed together. Each plant part has a different shelf life, activity, concentration, and consistency. Because of this, it is difficult to conclude stability of the final herbal formulation.

15.3.5 Drug Interactions, Deterioration, Decomposition, and Storage

Interactions between the active substances, microorganisms, and packing materials cause the majority of natural products to lose their effectiveness. Moisture is crucial for determining the growth of mold and fungus. Glycosides, alkaloids, tannins, and other active components of herbal products are stable under different conditions. Therefore, drawing any conclusions regarding the real stability of the product or formulation is challenging. Additionally, the herbal product can become toxic while being stored due to microbial byproducts, particularly mycotoxin.

15.4 DIFFERENT TECHNIQUES TO DEAL WITH INSTABILITY PROBLEMS RELATED TO HERBAL MEDICINES

15.4.1 Topical Herbal Formulation

The majority of herbal topical preparations are used to treat conditions including eczema, acne fistulas, piles, and other skin conditions. Today's cosmetic products, including face washes and sunscreen, often contain herbal ingredients—and most of the time, instability is evident. When the therapeutically active aqueous phase is integrated into the oil phase, a microemulsion or nanoemulsion formulation that increases stability and efficacy is produced. In addition, these formulations have a reduced impact on skin irritation and controlled delivery.

15.4.2 Suspension Form of Herbal Products

The most important factor affecting the suspension's stability is viscosity. Cellulose derivatives, natural polymers, gum-like xanthan, etc., are viscosity builders that stabilize the sparingly soluble or insoluble plant extracts.

15.4.3 Herbal Products in Emulsion Form

In the formulation, emulsifying agents are utilized to disseminate plant extract in the oil phase. The globule size and viscosity are evaluated during the entire storage investigation. In some cases, emulsifiers such as methylcellulose degrade and the viscosity decreases during storage, causing the emulsion to become unstable.

15.4.4 Natural Hair Color

Herbal-based natural hair coloring is a very popular formulation. Recent research has resulted in the development of a novel procedure and recipe for stable and secure hair coloring. The oil phase is included in the water phase of the formulation, together with another ingredient, such as an alkalizer, and is then further homogenized. It is crucial that a herbal hair dye is risk-free, allergy-free, mild in nature, does not harm hair, has a nice dyeing result, and has a glossy finish. It should also be free from contamination of the scalp and skin. Vegetable proteins are used in replacement of bleaching chemicals for making hair dye as emulsifying agents.

15.4.5 Herbal Preparations Including Significant Amounts of Extracts from Herbal Medicines

The desirable stability of preparations like tablets that contain a large number of herbs necessitates moisture adjustment. Other moisture-controlling chemicals—such as crystalline cellulose, croscarmellose sodium, magnesium stearate, and sodium dioctyl sulfosuccinate—are also used for this purpose. Calcium silicate is also used.

15.4.6 Semi-Solid Preparations Based on Supercritical Carbon Dioxide

Using the supercritical carbon dioxide process, plant extracts or essential oils produce a significant number of active principles that have been shown to be successful in wider stability experiments.

15.4.7 Liquid Formulations and Antioxidants

Antioxidants, esters of flavanol, and fatty acids are added to liquid herbal formulations of Glycyrrhiza, Ginkgo biloba, and polyphenols to reduce auto-oxidation.

15.4.8 Linctuses for Herbal Preparations

In general, linctuses are stable compositions on a physical level. Glycerin and hydroalcoholic extract used together give suitable stability by modifying the viscosity of compositions.

15.4.9 Plant Pigments

Natural coloring is preferred since it is less susceptible to photodegradation than plant pigments like chlorophyll, caramel, saffron yellow, and red beet pigments. In order to improve the stability of a compound, free radical production inhibitors.

15.4.10 Formulation of Tablets for Volatile Liquids

Coatings are used in tablet formulation to stabilize volatile oils. Additionally, moisturizing agents are needed to stabilize the tablet. Moisturizing agents such colloidal silicon dioxide and other materials are added before the entire tablet is mass-pulverized to a specific size to create dry granules. Before producing a tablet, additional excipients such as lubricants and disintegrating agents are mixed. Using a water-insoluble polymer, damp-proof coating is used for sealing or pre-coating, followed by sugar coating. Chewable herbal tablets made from Herba *Artemisia scopariae*, *Ardenia jasminoides* fruit, and *Lonicerae flos* are prepared utilizing excipients and exhibit high bioavailability and stability.

15.4.11 Chelating Agents for Aqueous Plant Extract Stabilization

In order to stabilize the herbal extract, most plant extracts are stabilized using a water-soluble chelating agent such as polyvinylpyrrolidone (PVP).

15.4.12 Powders Having an Oil Composition That Contains an Aqueous Active Ingredient

The flavor and plant extract oil constituents are disseminated in the powders, and the powder masks the herbal oils. The gum Arabic, (starch hydrolyzate), which has the consistency of powder, is used to improve the stability and cover up the taste of herbal oil formulations.

15.4.13 Vitamin Solutions

The addition of high-content glycerin, citric acid, and water via complete solubilization approach stabilizes and extends the half-life of vitamins and nutraceutical plant extracts.

15.5 TESTING THE STABILITY OF A FINISHED PHARMACEUTICAL PRODUCT (FPP)

The foundation for stability testing of a finished pharmaceutical product (FPP) is the pre-formulation study and the substance utilized in the pharmaceutical formulation. The stability of an FPP is affected by changes in the environment's temperature, humidity, and lighting. The shelf life or expiry limit, storage parameters, and in-use stability are only a few of the many determining factors of FPP that are developed for stability testing. Any finished product should have a shelf life parameter since it indicates the bio-efficacy and quality of FPP without compromising safety. Additionally, storage conditions and duration are indicated by stability data. The standard protocol is often followed when withdrawing samples from the FPP for stability testing, and it is appropriately documented. Typically, according to normal protocol, two samples are taken from the pilot-scale manufactured product, whereas one sample is taken from the production batch. In the pilot-sized batch of FPP, 10% of the pharmaceuticals' solid dosage scale, or 100,000—whichever is more—are withdrawn.

15.5.1 Evaluation

15.5.1.1 Organoleptic Characters

15.5.1.1.1 Appearance

The consistency and look of the products—including liquid and semi-liquid, as well as solid—are carefully monitored. Throughout stability testing, every modification to consistency or appearance is meticulously recorded and reported.

15.5.1.1.2 Color

Munsell color charts are usually used for the color determination of an FPP, and any alternation of color during stability testing is carefully reported.

15.5.1.1.3 Odor

The quality of the product can be assessed by considering the odor sensations. Changes in smell are a sign that a product is degrading, and it is simple to identify distinct smells like moldy, musty, fruity, fragrant, and rotten. The reconstruction of odor depends on individual perception, making it challenging to draw an exact conclusion. Changes in odor during stability testing are meticulously recorded.

15.5.1.1.4 Taste

Another crucial aspect of pharmaceuticals or herbal remedies is taste. Processed tea is assessed by trained tea tasters, and the pungency of capsaicin is determined by an expert assessor. Similar to odor identification, taste evaluation depends on individual perception, making it challenging to draw accurate conclusions. However, a sizable number of specialists or volunteers are enlisted to obtain accuracy. Typically, a professional tester has a limited social life, including a restricted diet. FPP is serially diluted in order to reach a conclusion. Changes in flavor—especially the intensity of flavor—are meticulously recorded during stability testing.

15.5.1.2 Physical Parameters

15.5.1.2.1 pH

The strength of 1% or 10% solution of FPP is evaluated by the standardized glass electrode.

15.5.1.2.2 Moisture Content

Moisture is a factor in product degradation due to microbial damage or when the active ingredients are hydrolyzed and lose their bioavailability (the azeotropic method or toluene distillation is typically used to measure moisture). In specialized equipment, drugs and toluene refluxed for six hours. Toluene is used to separate the drug's water content, and the percentage of moisture is then computed.

Moisture balance is also used for moisture determination on the basis of the weight loss after constant heating.

15.5.1.2.3 Ash Value

15.5.1.2.3.1 Total Ash After an herbal drug has been incinerated, the residual material—known as the ash content of the drug—is what remains and is what causes the product to deteriorate because of microbial damage or the active ingredients. Ash value is a measure to assess the identity and purity of the crude medication. The inorganic salts of phosphates, carbonates, sodium, potassium, magnesium, and calcium—mostly in oxide form—are found in ash-containing materials. The presence of mineral elements in plants is significant because they can sometimes act as pollutants or toxins that are harmful to human health, such as lead, mercury, and cadmium, but they can also have positive effects that are necessary for a balanced diet. The Indian Ayurvedic Pharmacopoeia now includes determination of As, Ca, Fe, Mg, Na, K, Zn, Ni, and Co, among other elements (The Ayurvedic Pharmacopoeia of India, 1999).

Using a tarred crucible, a known quantity of dried powdered medication (about 3–4 g) is incinerated at a temperature of 500°C to 600°C by progressively raising the temperature until it turns white, signifying that it is free of carbon. After the materials have cooled, the percentage of total ash is calculated using the air-dried sample as a reference.

The total ash content of a specific crude medication can be used to draw the following conclusions:

1. Quality of products.
2. Determination of exhausted products, e.g., mother clove and exhausted clove can be easily classified by determining ash content.
3. Detecting excess of sandy or earthy matter with drug.

15.5.1.2.3.2 Acid-Insoluble Ash Since the majority of earthy material is insoluble in acid, this approach can be used to check the quality of crude drug roots, rhizomes, and leaves. Most rhizomes and roots of crude drugs have calcium oxalate crystals, which serve as a quality indicator and can vary depending on the environment.

After filtration, the insoluble material is collected on ashless filter paper from the total acid-insoluble ash that had been added to 25 ml of diluted HCl. After thoroughly cleaning the filter paper with hot water, incinerate it at 450°C for 15 minutes, gradually raising the temperature while keeping it in desiccators. Acid-insoluble ash is measured in mg/g of the materials after cooling.

15.5.1.2.3.3 Water-Soluble Ash Some portion of earthy matter is soluble in water and marker of quality of root or rhizomes crude drug. Total ash is put into

25 ml of water and boiled for five minutes, followed by filtration by ashless filter paper. After washing the filter paper with hot water, incinerate at a temperature of 450°C for 15 minutes by gradually increasing the temperature. After cooling of the materials, acid-insoluble ash is calculated in mg/g of material.

15.5.1.2.3.4 Swelling Factor An essential indicator of the quality of a drug containing mucilage is the swelling factor. Long-term storage and impurity changed the swelling factor. A 25 ml stopper flask is filled with 20 ml of water and 1 g of the crude drug. The volume that has been occupied after 24 hours of standing and intermittent shaking is assessed and is known as the swelling factor.

15.5.2 Chemical Parameters

15.5.2.1 Extractive Values

This method is used for quality control, especially when standard procedures are unavailable or it is difficult to estimate the drug's contents using any other method. This technique aids in identifying adulterants and indicating the existence and kind of chemical ingredients.

15.5.2.2 Aqueous Extractive Value

Water-soluble extractive value has significance to the quality control of the drugs which contain water-soluble constituents such as tannins, sugars, plant acids, and mucilage. Macerate about 5 g of coarse powder of drug with 100 ml of water in a volumetric flask for 24 hours while shaking frequently. After filtration, evaporate about 25 ml of water extract to dryness in a tared flat-bottomed shallow dish followed by drying at 105°C for six hours then cooling in a desiccator; and weight is calculated in mg/g of material.

15.5.2.3 Alcohol Extractive Value

Asafetida and myrrh are officially analyzed using this approach. Alcohol-soluble extractive value is important for quality control of pharmaceuticals that contain alcohol-soluble ingredients such as tannins, resins, and alkaloids.

In a 100 ml stopper flask, macerate 5 g of coarse drug powder with 100 ml of 90% alcohol for 24 hours while stirring frequently. After filtering, around 25 ml of alcohol extract is evaporated to dryness in a flat-bottomed shallow dish coated with tar. This is followed by drying at 105°C for six hours, cooling in desiccators, and calculating weight in mg/g of material.

15.5.2.3.1 Ether Extractive Values

Alcohol-soluble extractive value has significance to quality control to the drugs of volatile oils, fixed oils, and resins. Ether extractive values are determined as described earlier.

15.5.2.4 Tannins

Estimation of tannin has significance to the quality control of the drugs containing tannins. Tannin content is determined by simple hydrolyzing with water and gravimetrically estimated.

15.5.2.5 Total Alkaloids

A simple perception reaction with ammonium hydroxide is used to determine the majority of herbal drugs' alkaloids. In a conical flask, 4–5 g of the drug are taken along with 200 ml of 10% acetic acid in ethanol. After the mixture has concentrated in a water bath for four hours to about one-fourth of its initial volume, concentrated ammonium hydroxide is added to precipitate the aluminum salt of the alkaloids. Alkaloids are recovered after thorough filtering, then they are dried and weighed.

15.5.2.6 Glycosides

Glycoside is reactive with Fehling's solution. About 10 g of drug with 100 ml of 50% sulfuric acid (H_2SO_4) is taken in a conical flask and heated in boiling water for 15 minutes followed by 10 ml of Fehling solution added, and further the mixture is heated to obtain a red precipitate. After filtration, the red precipitate is collected and the percentage glycoside is calculated.

15.5.2.7 HPTLC Fingerprinting

One of the critical parts of determining the active components in herbal medicines is HPTLC. Chemically defined components known as markers are in charge of therapeutic effects. The quality of the finished product, determined by evaluating of marker in the herbal drug or preparation, has been quantitatively determined.

15.5.2.8 Microbiological Assay

One of the critical parts of determining the active components in herbal medicines is HPTLC. Markers are in charge of therapeutic effects. To assess the quality of the finished product, a marker in the herbal drug or preparation has been quantitatively determined.

15.5.2.9 Microbiological Assay

The microbial load of the end product can occasionally be potentially toxic to humans. Microbe contamination during the production and preparation of herbal medication is inevitable. Temperature and humidity during storage affect microbial infection. Storage and the growth of molds and fungi are accelerated because there is no possibility of heat exchange. The majority of the time, mycotoxins—which are toxins produced by fungi—are harmful to humans. The WHO procedure is the one that is most frequently used to figure out the total number of microorganisms in plant materials. Authorized monograms, viz., pharmacopoeias from the United States or India, advocated for accepted evaluation techniques. In accordance with WHO guidelines, 90 ml of buffer sodium chloride-peptone and 10 g of sample are taken (pH up to 7.0). By planting depth on casein-soybean digest agar, followed by incubation at 30–35°C for 48 hours, total aerobic bacteria are counted in duplicate while Sabouraud dextrose media and 10% tartaric acid (pH 3.0–3.5) are used for determination of total yeast and mold and colony counted in duplicate after incubation at 20–25°C for five days. Other particular pathogens, such as those in the family Enterobacteriaceae and other gram-negative bacteria (such as *E. coli*, *Salmonella sp.*, *P. aeruginosa*, and *S. aureus*), are identified using particular culture techniques,

TABLE 15.1
Standard Microbial Limits in CFU/g

Authorities	Materials	Aerobic Bacteria	Mold and Yeast	Enterobacteria and Another Gram-Negative Bacteria	E. coli	Salmonella
WHO	"Crude" plant material	-	10^4	10^4	10^4	Absent
	Topical dosage forms with boiling water	10^7	10^3	10^3	10^3	Absent
	Other plant materials for internal use	10^5	10^2	10^2	10^2	Absent
USP	Dried or powdered botanicals without boiling water	10^5	10^3	10^3	absent	Absent
	Tinctures, powdered botanicals extracts, fluid extracts, and nutritional supplements with botanicals	10^4	10^2	-	absent	Absent
	Infusions/ decoctions	10^2	10^1	-	absent	Absent
EP	Herbal medicinal products with boiling water added before use	10^7	10^5	-	10^3	Absent
	Herbal medicinal products without boiling water added before use	10^5	10^4	10^3	absent	Absent

as well as other approaches including biochemical and serological assays, used simultaneously. WHO standards for acceptance for yeasts and molds are 10^4 colony forming units (CFU)/g as infusions and 10^3 CFU/g for internal use, whereas WHO limits for total aerobic microorganisms are not more than 10^7 CFU/g for plant material in infusions and 10^5 CFU/g for internal use like suspension, tablets, etc. The possibility of producing a mycotoxin, which can cause cancer, such as aflatoxin, has led to significant fungal counts being regarded as a risk to human health, as previously stated.

15.6 PROTOCOLS OF STABILITY TESTING OF DRUG SUBSTANCES/ACTIVE PHARMACEUTICAL INGREDIENTS

15.6.1 Selection of Batches

Data is obtained after the selection of a minimum of three first batches of the pharmaceutical component. The pilot process variability of the three designated batches should be comparable, including the synthetic route, manufacturing technique, and every other operation that is intended to serve as a prototype for the actual production process. Additionally, it is intended that the chosen batches serve as a good representation of the final product.

15.6.2 Container Closure or Final Packaging

It is recommended that research be done in the claimed commercial packaging (container-closure) or mimicked or stimulated packaging conditions.

15.6.3 Guidelines or Specifications

The specific processes or protocols of stability testing—comprising a list of tests, analytical procedures, recommended acceptance criteria, and related references—were directed by documents Q3A, Q6A, and Q6B of the ICH guidelines. A stability study provided the characteristics and suggestions of the potential bioavailability or potency of drugs during storage, as well as the influence of quality, safety, and the possible incorporation of such criteria. The study should be comprehensive and take into account characteristics that are physical, chemical, biological, and microbiological. The analytical process should be validated simultaneously with the stability study.

15.6.4 Storage Conditions

Storage requirements and testing frequency must be satisfied according to the specified ICH Q1A (R2) standards. Drug substance stability is primarily assessed in terms of two factors: susceptibility to moisture and impact of storage temperature. As was previously noted, herbal medicines are more susceptible to moisture and temperature, which promotes microbial growth. The design of the storage conditions takes into account transportation, shelf-life claims, and warehouse storage.

At the time of submission for regulatory clearance, the long-term testing is carried out in three main batches for a minimum of 12 months. Because perfect storage conditions cannot be anticipated, accelerated storage conditions offer the conclusive evidence of short-term occurrences, which are mostly important in transportation. According to ICH protocols, accelerated studies are carried out at 40 ± 2°C and 75 ± 5% RH for six months, while long-term stability testing is performed at 30 ± 2°C and 65% ± 5 RH for 12 months, with some exceptions made in accordance with Supplement 2 to the Main Generic Guideline. Supplement 2 was used in accordance with ICH protocol for the evaluation of data conditions for refrigerated storage. The experiment refers to real-time data collection at long-term storage conditions when any discernible changes occur from 3–6 months at the accelerated storage condition.

If any substantial changes occur during accelerated stability testing within three months, the impacts of shipping, handling, or brief excursions outside the label storage conditions should be addressed.

The discussion provides justification for conducting more testing in one batch for a shorter time frame than three months but more frequently than normal. In this scenario, a six-month study period is not important because there have been noticeable changes within three months or less. According to ICH protocols, accelerated tests are carried out at 30 ± 2°C and 65 ± 5% RH for six months, while long-term stability testing of chilled storage is performed at 5 ± 3 °C for 12 months without defined RH conditions.

Additionally, ICH advised producing the short-term data following freezing. Prior to being intended for storage in a freezer, real-time data at the long-term storage state must be obtained. In the absence of accelerated storage data, medications are kept in a freezer to prevent temperature impacts during handling, shipment, and short-term excursions outside the label storage conditions. The suggested temperatures are 5 ± 3°C or 30 ± 2°C for the appropriate amount of time based on a single batch study and prerequisites. Additionally, the ICH advised case-by-case drug-testing below –20°C and in accordance with intended use of the geo-climatic circumstances.

15.6.5 Data Interpretation and Evaluation

After analyzing degradation of products or original content, a shelf-life variation limit of 10% from the label claim is permissible. The ICH Q1A (R2) protocol is followed for both storage conditions and testing frequency. The suggested storage conditions for prequalification or pre-registration requirements, unless otherwise justified, are 30 ± 2°C and 65 ± 5% RH, with "substantial change" being interpreted in accordance with ICH Q1A (R2) protocol. There are places to relax throughout the registration or approval process. It is necessary to provide assurance or a promise to continue the stability studies after approval and in compliance with the proposed retest period because the long-term stability data on primary batches do not comply with it. A post-approval commitment is taken into consideration in accordance with the submission, which includes long-term stability data on three production batches encompassing the suggested in retest period. As an alternative, if fewer than three batches of long-term stability are provided, a promise can be made to continue producing more batches over the suggested retest time. Unless there is a scientific justification for doing otherwise, the long-term stability technique should be replicated for primary batches. After conducting stability studies, sufficient analytical data should be produced, including tests for chemical and antioxidant preservatives, bio-efficacy, and other quality indicators such microbiological contamination, sterility, or the presence of bacterial endotoxins.

15.6.6 Validation

The level of confidence for accepting the data should be within 95% of the variability of individual batches and future production batches within the given retest period. The statistical method is also used to check the data, and there is usually not much difference between the data submitted and the data in the next batch. This makes it unnecessary to do a formal statistical analysis, but a good reason should be given for not doing so. Even though there is not much variation from batch to batch, the data were put together to get a single overall estimate. This estimate was based on statistical tests and the significance of "p" values > 0.25. Slopes of each batch's data is analyzed by drawing regression lines and zero-time intercepts. With linear regression analysis, find out what kind of relationship there is between degradation and storage time. Depending on how stable the drugs are, a linear, quadratic, or cubic relationship could be drawn, and further, from the function, an equation on an arithmetic or logarithmic scale could be derived to curve fit or regression equation fit with proper statistical analysis of the data on all batches and the combined batches. Extrapolation of the data is a very important part of long-term stability or accelerated stability studies. If enough justification is given, real-time data from long-term storage can be used to extend the retest period, even if only a small amount of data is submitted. Justification should include the mechanism of degradation, a supported mathematical model, the conditions of accelerated studies, or other supporting data like batch size or stability data. Also, extrapolation data show what the degradation product is and how it should be attributed.

15.6.6.1 Disclosure

The label on the container should include a statement of how it should be stored. This is a legal requirement. "Ambient conditions" or "room temperature" should not be used in this context.

15.6.7 Finished Final Formulation (FPP) or Drug Product (DP)

Stability studies are designed in such a way as to correspond with the product cycles of the finalized product. In the developing phase and the registration phase, the planning and protocols of stability studies are carried out in different ways.

15.6.7.1 Developing Phase

Accelerated stability studies are short experiments that are used to evaluate new developments in formulations, packaging materials, and/or manufacturing processes. The goal of these investigations is to gain an idea of the product that has been developed. After the optimization of the formulations, the accelerated studies examine whether or not the stability can be accurately predicted, whether or not the optimal storage conditions are optimized, and whether or not the expiration date can be accurately forecasted. When conducting stability studies, it is essential to take into account multi-dose containers as a further crucial component. Real-time studies should be done for the purpose of confirming the substance's stability, and these studies should begin as soon as possible after the final formulations have been determined.

15.6.7.2 Registration Phase

The manufacturer is expected to submit information and data concerning the stability of the product's final container and packaging, in addition to the results of any relevant tests for the registration of the final product. At the time of final registration, both accelerated and real-time data must be submitted. The requirements of stability and storage that are described "in term" should be included on the label of the reconstituted dry powder that will be used for injecting patients or making oral suspensions. At the time of registration, established or experimental supplementary data of storage, stability, or formulation data are also considered considerable data. At the time of registration, you can be asked to submit the data on the stability of the initial production batches.

15.6.7.3 Post-Registration Phase

A crucial step in the life cycle of a product happens after it has been registered. Real-time stability data are required to be generated in order to confirm the anticipated expiration date and projected storage conditions that were provided during the registration process. In the post-registration step, concerns regarding the safety of degraded goods are also taken into account. The details of the stability data and validation are carefully examined during the good manufacturing practice (GMP) approval process. Any modification to the formulas, the manufacturing process, or the packaging must be reported and counted as supplemental data.

15.6.7.4 Variations

After the registration of the finished product, new stability studies must be carried out if there is any major change made to the product. The investigator is obligated to report the data to the relevant regulatory body for the purpose of ensuring that a change in the formulations does not affect the qualities of the active ingredients and/or pharmaceutical goods that are being changed by the product. Studies on stability are planned such that they take into account the active elements. Accelerated studies are carried out whenever there is a change in either pharmaceutical product's composition or its production procedure, as well as whenever there is a change in the immediate packaging.

15.6.7.5 Design of Stability Testing

The fundamental idea behind the design of stability studies for FPP is that they should be based on the qualities of the active ingredient, which may include its physical or chemical properties. Any FPP worth its salt will take into account climate as an essential component of its planning. Ointments made from herbs are more likely to deteriorate in environments that are hot and humid. According to the World Health Organization (WHO), the world is divided into five major zones for the purpose of conducting stability studies.

More preciously, the stability studies are dependent on the mean kinetic temperature. Changes in temperature, humidity, light intensity, or partial vapor pressure can all have an effect on stability studies. Both semi-solid and liquid dosage forms are prone to degradation at low temperatures, so it is important to test their stability at temperatures ranging from 0–−20°C (freezer), as well as at freeze–thaw cycles and temperatures ranging from 2–8°C.

Some herbal products have been coated with polymer or other materials that are USFDA-approved, and these products are required to undergo stability testing when

TABLE 15.2
Zones of Stability Studies of Herbal Pharmaceuticals

Zone I	Temperate
Zone II	Sub-tropical with possible high humidity
Zone III	Hot and dry
Zone IVa	Hot and humid
Zone IVb	Hot and very humid

exposed to light. Every photostability test imaginable was performed on at least one primary batch of the product.

15.6.7.6 Selection of Batches

Stability data are necessary for a minimum of three primary batches of the same formulation and dosage form in the container and closure as planned for the marketing of the marketable FPP goods in order to register them with the regulatory authorities. Among the three batches that are specified, a minimum of two of the batches should be pilot scale, and the other batch may be large as a smaller batch. In the final manufacture, each batch of FPP must be similar and easily recognized in accordance with the specifications. Bracketing is used instead of the alternative, which results in a more accurate and valuable result matrix design of stability experiments done on each individual strength and container size of the FPP.

15.6.7.7 Packaging/Containers

Stability tests must be performed on the FPP packaging and containers that are being considered for commercialization. It is necessary to analyze whether or not the drug or active ingredient interacts with the packaging material, which includes both primary and secondary packaging. Additionally, accelerated stability testing is helpful for evaluating the effects of stress on unprotected pharmaceuticals, as well as the changes in storage conditions that occur during transit. This type of testing is considered to be an important component of stress testing during the product development cycle.

15.6.7.8 Storage Conditions

An FPP is put through a series of tests to determine whether or not the sample demonstrated favorable thermal stability, susceptibility to moisture, and the possibility of solvent evaporation. Stability tests take into account a number of significant circumstances, including multiple-dose containers, subsequent use, and shipment. If the product is intended for use or has been reconstituted, the storage conditions need to be indicated on the label or in the accompanying leaflets. Illustrations of such goods are oral suspension, powder, and parenteral preparation. When conducting stability tests on this kind of product, great care is needed to take into account the efficacy and stability of the product after it has been diluted or reconstituted, and stability tests are carried out under reconstituted conditions both at the beginning and the end of the process. Long-term research is carried out in order to arrive at conclusions regarding the product's stability, as well as its shelf life. Long-term stability tests are carried out for a minimum of a year on each of three batches, as previously mentioned; these tests ought to be sustained for actual suggested self-studies. As the conclusive consequences of short-term excursions beyond the storage conditions indicated in labels, which come into effect at the time of shipping, it is necessary for accelerated storage conditions to be in place. In the event that FPP is not covered by the section, the general storage conditions should be reported to the subsequent section, and the appropriate justification of storage conditions are to be indicated. At the time that the suggested shelf life is being considered, data on the generic product's long-term stability are necessary.

Medicinal products sometimes become unstable under accelerated settings and significant change results—as a result, data cannot be collected; in such a scenario, extra testing is required at the intermediate storage condition.

When and if long stability studies fail at the storage conditions mentioned previously, the FPP are considered in the reduced shelf-life claim. Alternatively, a more protective container closure system should be used, along with additional cautionary statements in the labeling of predictive causes of degradations. The data on the material's stability in chilled conditions was also utilized as the material that was taken into consideration for use in various climatic zones, including those with colder temperatures, taking into account that the transition from warm to cold weather in India is fairly sudden during the winter season. Accelerated storage tests that last 3–6 months are typically recommended. If any product fails the study, the stability of the product can be deduced from the real-time data on the product's state while it is stored for a lengthy period of time. In addition, the short time-accelerated study conducted over the course of three months was unsuccessful; more data should be supplied to cover short time excursions, including storage throughout the time of shipping and handling. In light of the fact that the known FPP stability did not meet the recognized criteria within the first three months or afterward, a study that lasted for six months would have been useless. In spite of this, a single batch study is recommended instead of three batches of FPP, each of which has a stability of less than three months when subjected to accelerated settings. The information from one batch of FPP products kept in a freezer or normal condition and data from 5°C ± 3°C or 30°C ± 2°C at an appropriate time period to conclude the effect of short-term

TABLE 15.3
Storage Conditions of a Finished Pharmaceutical Product

Conditions Study	Condition of the Study	Minimum Time Period Covered
Long-term	30°C ± 2°C/65% ± 5% RH	Twelve months
Intermediate	30° ±2°C/65 ±5% RH	Six months
Accelerated	40°C ± 2°C/75% ± 5% RH	Six months

TABLE 15.4
Storage Conditions of the Finished Pharmaceutical Product

Conditions Study	Condition of the Study	Minimum Time Period Covered
Long-term	5°C ± 3°C	Twelve months
Accelerated	30°C ± 2°C/65 ±5% RH	Six months

excursions at the time of proposed label storage condition are sometimes used as an alternative to accelerated storage conditions.

In addition, the FPP are kept at a storage temperature of less than –20°C, which is a factor that has a significant impact on the conditions of the simulation while it is being used.

15.6.7.9 Elevated Temperature and/or Extremes of Humidity

These data are a supplementary or additional requirement of the categorized product that is used in such climatic conditions or transportation and climatic conditions that are outside of the storage conditions. When conducting stability experiments on water-vapor permeable packaging material, such as PVC/aluminum blisters for solid dose preparation, the high humidity condition (for example, 30°C with 80% RH) is taken into consideration. Also, the FPP is likely to be marketed in Zone IVa, which indicates that testing should take place in regions that have conditions characterized by exceptionally high humidity levels. In this particular scenario, the packaging of primary containers with aluminum or aluminum blisters or another barrier to water vapor is not considered necessary for stability testing when the relative humidity is exceptionally high.

15.6.7.10 Determining Moisture Loss

In stability studies, moisture is a very important component to consider. The presence of an excessive amount of moisture causes the items to become brittle, and it allows microbes to contaminate herbal products. The efficacies of the medication, as well as its structural integrity, are both compromised when the solid dosage form of FPP is exposed to moisture. The final container and its closure system have a significant role to play in preventing either moisture loss or moisture gain. The following formula can be used to determine the ratio of water loss that has occurred:

$$\text{Change in Moisture Content (\%)} = [(W2 - W1)/W1] \times 100$$

Where:

W1 is the initial moisture content (before exposure).

W2 is the final moisture content (after exposure).

TABLE 15.5
Water Loss Rate Ratios at Relative Humidity Conditions

Relative Humidity	Reference Relative Humidity	Water Loss Ratio Given Temperature
60% RH	25% RH	1.9
60% RH	40% RH	1.5
65% RH	35% RH	1.9
75% RH	25% RH	3.0

This formula helps in quantifying the change in moisture content as a percentage of the initial moisture content, which is essential in assessing the impact of moisture exposure on the product's stability.

$$(100\% \text{ reference humidity}/100\% \text{ relative humidity})$$

Throughout the storage time, a linear data set representing the water loss rate at the alternate relative humidity.

15.6.7.11 Microbial Testing

Microbiological contamination, which was the source of the instability of both sterile and non-sterile multiple-dose FPP, must be addressed. Preservatives are most commonly used in order to prevent microbiological contamination, and the utilization of preservatives is indeed very strategic in three different regards. First, we assess the safety aspects of the preservative, then we look at the content of the preservative, and finally, we establish the efficacy of the preservative by subjecting it to a series of challenging tests. The integrity of the dose form is evaluated after the expiration of the product's shelf life, and the content of the preservative and active component is reported accordingly.

15.6.7.12 Bacterial Endotoxins and Pyrogen Testing

Pyrogens and bacterial endotoxins should not be present in sterile dosage forms such as large volume parenteral, small volume parenteral, or forms comprising dry materials (powder-filled or lyophilized goods) or solutions. During the period of stability, it is important to conduct adequate tests of the parameters at acceptable intervals.

15.6.7.13 Analytical Methods

At the time of the stability testing, the characteristics of the FPP that are likely to be affected are evaluated, and likewise, after the storage of the FPP, the potency, content, or decomposition of the active component, as well as the interaction with the containers, is evaluated. At the time of the stability research, the various physical parameters, such as hardness, dissolution, particle size, disintegration, turbidity, and particulate matter, are examined. It is required to demonstrate the efficacy of additives, anti-microbial agents, and preservatives in order to keep the developed product's integrity intact, several testing procedures need to be devised and validated.

15.6.7.14 Testing Frequency

It is necessary to conduct frequent testing due to the considerable data integration and the formulation of stability characteristics. Long-term testing should ideally be performed once three times a year in the first year, six times a year in the second, and annually afterwards. At least three time points, including the beginning and end of the experiment, are used to conduct accelerated stability tests (e.g., zero, three, and six months). When tests at the intermediate storage state are necessary experiment of the accelerated condition of storages required and it is recommended to use a minimum of four time points, including the initial and final time points (for example, zero, six, nine, and twelve months) from a study that lasts for a year. If the expedited studies yielded extraordinary results, including major changes in criteria and the application of testing techniques, then more samples should be included at the appropriate time points in the protocol (e.g., zero, one, two, three, and six months). It is possible to choose either the matrixing or the bracketing approach in order to reduce the number of study protocols.

15.7 DEFINITION OF "ACCEPTANCE CRITERIA"

The stability test has to cover any relevant parameters, such as quality, safety, and/or efficacy, which are vulnerable to change or affect while the product is being stored. The physical, chemical, biological, and microbiological preservative content are directly attributed by the stability data and the product's characterization. The analytical processes should be properly validated, and stability indicating shelf-life acceptability criteria should be determined from taking into consideration all of the available stability information. The justification and applicable solution should be there about the stability evaluation and the changes detected in storage and between correlations. Also included in the required information is that regarding the stability data evaluation. The anti-microbial or preservative or stabilizer or antioxidant used in the formulation should be documented, as well as the particular utilizing content that needed to be justifiable for development data. Additionally, any changes to the shelf life and release data should be reported. In most cases, a single original batch is used for testing to determine whether a preservative, stabilizer, or antioxidant is effective.

15.8 DEFINITION OF "SUBSTANTIAL CHANGES"

A "substantial change" is defined as an alteration in the assay that is greater than or equal to 5% of its initial value. Changes in appearance, color, phase separation, re-suspensibility, hardness, and cracking are some examples of the kinds of physical changes that can be attributed to the conclusion that "substantial changes" have occurred. Temperature can modify the look as well as the viscosity of semi-solid formulations such as creams, ointments, and suppositories. The re-suspendability decreased at low temperatures, but increased at high temperatures; this phenomenon was also noticed at the time of accelerated research. Softening, melting, and sometimes leaking are observed during the summer season or at higher temperatures.

Stability tests for multi-dose containers and formulations typically involve many different doses of the drug. The consistency of the contents of multi-dose containers was compromised when they were opened and closed multiple times. In the test protocol, there must be a minimum of three batches evaluated; at least two of these batches must be conducted on a pilot scale, and one of these batches must have completed its real-time stability tests and be closer to the product's expiration date. Because impermeable containers lose the least amount of moisture and allow tests to be replicated in either controlled or natural levels of humidity, they are considered to be more acceptable. During the course of the stability investigation, it is necessary to detect any degraded products.

15.9 EVALUATION OF THE RESULTS

Methodical presentation and evaluation of the stability study data includes consideration of all parameters, including physical, chemical, biological, and microbiological tests, as well as other physical characteristics, such as dissolution rate, disintegration, and hardness tests. Storage requirements are relevant to all upcoming batches once the stability of three batches has been established, as previously stated. Additionally, the same conditions should be kept in place for the manufacture and packaging of drugs in the future. In extremely modest variations of the stability studies, statistical analysis is superfluous and omitted with the appropriate explanations. The declared shelf life should be met with linearity of confidence at one-sided 95% and variability of individual production batches. Additionally, when batch-to-batch variability in statistical data is minor, it is more appropriate to integrate the data into the final resultant by applying the proper statistical techniques to calculate the proper "p" values for level of significance when employing regression analysis, where the conclusion is based on the slopes of the regression lines and the zero time intercepts for each of the individual batches. The linear equation used in statistical analysis should have an arithmetic or logarithmic scale and be linear, quadratic, or cubic. It is required to ensure that if the shelf life falls below acceptable limits, statistical methods should be fitted with combined batches of numerous batches.

15.10 EXPLORATION OF DATA

By investigating the stability data from the accelerated studies, the conclusion may be drawn that there has not been a major change of the FPP's shelf life that extends beyond the time of the investigated long-term data. The basis for validating the extrapolation of stability data is the fitting of the given data into a mathematical model using a logical approach and knowledge, and the same mathematical model is used for future manufacturing batches. According to the long-term stability data, the products' evolving stability and degradation pathway should be followed linearly. The accuracy of the studied data is crucial, and when generating a regression curve or state line equation, the created data serve as the corrections by themselves. The shelf life of the drug is calculated using an analysis of long-term data, and additional data should be given when submitting to regulatory agencies. The type of the deteriorated product and its attribution to the stability were taken into account while evaluating the FPP's shelf life. The period of time measured from the batch's release to the market is its shelf life or expiration. The batch must be released within 30 days of the production date—if it is not, shelf-life tests are conducted again while taking into account the active component and the other ingredients.

15.11 DRUG PRODUCT STABILITY WHEN KEPT IN SEALED CONTAINERS

FPP is stored in impermeable containers since it is susceptible to solvent loss and is sensitive to moisture. A moisture barrier is provided by an impermeable container. In situations when solvent loss and concentration changes are a problem, aluminum tubes are used for semi-solid preparation and sealed ampoules are used for injectables. Studies on product stability for impermeable containers should be conducted in humid environments that are either controlled or ambient.

15.12 DRUGS FOR STABILITY PACKED IN SEMI-PERMEABLE CONTAINERS

The majority of FPPs are aqueous-based and are kept in semi-permeable containers, such as semi-permeable plastic bags, large volume parenteral (LVPs), small volume parenteral (SVPs), nasal drops, semi-rigid plastic containers, sealed plastic ampoules, and ophthalmic vials. These containers have the potential to lose water, so stability tests are carried out at low relative humidity levels to make sure that they are suitable for low humidity environments. Studies on the stability of FPP in semi-permeable containers also assessed its stability on physical, chemical, biological, and microbiological levels.

When kept in low relative humidity situations, semi-permeable membrane storage for aqueous-based FPP has a shelf life that is appropriately stated. If any FPP loss in the water of 5% within three months at 40°C/NMT 25%RH is deemed significant, then a major change might be taken into account. Additionally, the expedited stability studies were created with a purpose and real-time data, and they were used for the new product.

Stability studies with semi-permeable containers also reflected the transition periods of less than three months when the container stored to outside storage conditions

TABLE. 15.6
The Conditions for the Stability Testing of FPP with Semi-Permeable Containers

	Storage Condition	Minimum Study Period
Long-term	30°C ± 2°C/35% RH ± 5% RH	Twelve months
Accelerated	40°C ± 2°C	Six months

during the shipment, and such studies of single batches are carried out at the accelerated conditions.

15.13 LEGAL STATEMENTS AND LABELING RELATED TO STABILITY

A storage statement is a crucial component of the FPP and is therefore created or written in accordance with the regulations set out by various regulatory bodies, with shelf-life claims made in accordance with the findings of stability results. The expiration date was calculated based on storage circumstances and indicated on the label appropriately. It should have specific directions like "avoid sunlight", "do not store in refrigerator", "keep in refrigerator", or "shake well before use". Avoid using words like "ambient conditions" or "room temperature" that are ambiguous or unclear.

The label does not include any cautionary remarks under the purview of stability studies. The phrase "precautionary" may, however, occasionally appear on the label. For a reconstituted FPP of an antibiotic injectable or oral suspension made from antibiotic powder, the reconstitution process and the conditions under which the product was produced should all be mentioned.

15.14 ON-GOING/IN-USE STABILITY

Once the container is opened, the first dose is removed and the stability data are validated for in-use stability testing of multi-dose products at the end of the claimed shelf life. At least two batches from the pilot scale should have been withdrawn for testing, with one of those batches undergoing real-life testing. The investigation

TABLE 15.7
Legal Statement of the Storage of Containers

Storage Conditions	Storage Statement in Label
Room temperature	"Store below 30°C" or "Store below 25°C" if deemed essential in some cases.
Refrigerator	"Store in refrigerator, between 2°C and 8°C".
Freezer	"Store in freezer between –5°C and –20°C".

TABLE 15.8
Legal Statement of the Storage of Containers

Stability Problem	Precautionary Statement Which the Label Should State
FPP cannot tolerate refrigerating	"Do not refrigerate"
FPP cannot tolerate freezing	"Do not freeze"
FPP light sensitive drug products	"Store in original pack to protect from light"
FPP sensitive to humidity	"Store in a dry place"

should be encouraged in such a way that filing, volume adjustments, or reconstitution are considered before final usage. During in-use stability investigations, potential breakdown or degradation pathways are identified while taking into account physical, chemical, and microbiological factors. Specific conditions, such as the addition of a preservative or stabilizer, are also considered when the FPP is reconstituted, as is the suggested stability after opening the container. Normally, eye drops and reconstituted suspension are considered for three months. All marketed and licensed products must go through a yearly product quality review (PQR) in accordance with GMP rules to update, add to, or remove the consistency and validity of the entire manufacturing process, including stability studies. Stability testing differs from other stability tests in terms of regulatory and legal requirements. The PQR covers products containing herbal medicines, and when India required implementation of GMP, the continuous stability of herbals was thoroughly and unrestrictedly studied.

15.15 REAL-TIME SHELF-LIFE DATA OBTAINED THROUGH BRACKETING AND MATRIXING

The assigned time period during which the FPP demonstrates bioavailability within the specified time period is known as shelf life. When the degradation of active ingredient reaches 10%, it is considered an unstable point. A 24-month shelf life for a product is ideal, with shelf-life data starting from the date of manufacture. Shelf-life data are derived from accelerated studies and extrapolated with assistance from real-stability data. However, real-time studies, which must be submitted at the time of registration and continue until the suggested shelf life is achieved.

15.15.1 Bracketing and Matrixing

These are mathematical and statistical models that can reduce the number of test points with acceptable attributes proof of stability.

15.15.1.2 Bracketing

Bracketing represents the data of stability schedules in which extreme parameters such as container size and/or dosage strengths are tested at any time. The bracketing design can be extrapolated to the assumption of intermediate stability by representing extremes.

A new drug product that comes in three different strengths, two different packaging materials, and three different pack sizes for one of these materials can be evaluated with this method.

15.15.1.3 Matrixing

Matrixing represents the statistical design of a stability schedule in the process of comparing numerous points, including the considerable variation in the ratio of ingredients; modification of one or two minor components; use of different interface materials or closing systems (non-equivalent performance); production procedure; manufacturing site (another company) at any designated sampling point.

TABLE 15.9
Typical Bracketing Methodology of Stability Studies

Pack Type		Dosage Strength	
Strengths	50 mg	100 mg	200 mg
Batch	A B C	A B C	A B C
Laminated	z z z	(x) (x) (x)	z z z
LDPE/150	z z z	(x) (x) (x)	x x x
LDPE/200	(x) (x) (x)	(x) (x) (x)	(x) (x) (x)
LDPE/500	z z z	(x) (x) (x)	z z z

However, additional sampling points, the alternation of different sets of samples, and the total number of samples can be performed, and the design can be formed in such a way that the consistency of the batch represents the uniformity of all of the samples. In addition to that, this is a conceptual methodology that reduces testing plans in an acceptable manner so that the amount of time necessary for operations is decreased. It is important that the stability of the API and the end product be very similar, and this similarity should be well justified. Additionally, this concept ought to be examined further in terms of its acceptability to other products in the same category. As a result, the length of the study was substantially shortened, which is a phenomenon known as matrixing, whereas bracketing is restricted to closely limiting batch variability and reducing the overall experiment duration across all batches as recommended.

15.16 STABILITY OF HERBAL PRODUCTS

The nature of herbal medicinal products makes them complicated, and occasionally their formulations involve multiple ingredients, which might influence both their stability and their production process. Degradation and transformation into another derived product can take place in the finished herbal product without causing any reduction in the therapeutic response; this allows the finished herbal product to be considered a natural product. One of the examples is the process of making valerian roots, in which the active ingredient acetoxyvalerenic acid is first hydroxylated to produce hydroxyvalerenic acid, and then, after decarboxylation, it transforms into valerenic acid. This entire phenomenon is occurs throughout the manufacturing process without causing any reduction in the drug's effectiveness. Therefore, it is appropriate to include the "sum of valerenic acids" in the specification, and doing so is advised for quality control. Although the studies of the finished herbal product's stability are comparable to those of similar products, intermediate and accelerated stability assessments are not conducted. The majority of herbal products are unchanged at temperatures of 30°C and 65% RH, and in particular at temperatures of 40°C and 75% RH stability determined. Nevertheless, given sufficient justification, the regulatory authority can be acknowledged, but the labeling must still be done appropriately. Because different herbs are harvested at different times, combination products make up the vast majority of herbal goods, making it extremely challenging to give statistics on the relative stability of two batches of herbal goods. Because of the varying harvesting times, it is nearly impossible to produce the two batches of each extract that are required at the same time. This issue has been taken into consideration. Additionally, one of the two batches of the finished product will be selected at random to undergo real-stability testing at the end of the product's shelf life. When necessary, skip testing for various packaging sizes can be accomplished with the help of bracketing and matrixing studies. Because all of the preparations are admixtures of liquid extracts, it is required that data on the herbal liquid preparations' stability be collected while they are being used. Products made from herbs should be kept in the appropriate conditions and labeled.

BIBLIOGRAPHY

1. Bosch WH, Hilborn MR, Hovey DC, Kline LJ, Lee RW, Pruitt JD, et al. Liquid dosage compositions of stable nanoparticulate drugs have the improved stability than conventional dosage forms. *PCT Int Appl.* 2004:68.
2. Anna RB, Bergonzi CM, Giovanni M, Franco FV. Development and stability of semisolid preparations based on a supercritical CO_2 Arnica extract. *J Pharmaceut Biomed Anal.* 2006; 41:449–454.
3. Salunkhe V, Bhise SB. Formulation and stability studies of chewable herbal tablets. *Indian Pharma.* 2009; 8:53–58.
4. Xiangwu Y, Fan J, Yubin Z, Ting P, Kun Y, Wenchao Y. New chewable tablet formulation of mixture of Herba Artemisiae Scopariae, Gardenia jasminoides, and Flos Lonicerae. *Zhuanli Shenqing Gongkai Shuomingshu.* 2007:25.
5. Chaudhary M, Naithani V. Novel herbal topical formulations for treatment of acne and skin disorders. *Indian Pat Appl Publ.* 2009:28.
6. Beneke CE, Viljoen AM, Hamman JH. Polymeric plant-derived excipients in drug delivery. *Molecules.* 2009;14:2602–20.
7. Aziz NH, Youssef YA, El-Fouly MZ, Moussa LA. Contamination of some common medicinal plant samples and spices by fungi and their mycotoxins. *Bot Bull Acad Sin.* 1998; 39:279–285.
8. Mandeel QA. Fungal contamination of some imported spices. *Mycopathologia.* 2005; 159:291–298.
9. Gonda S, Tóth L, Gyémánt G, Braun M, Emrid T, Vasas G. Effect of high relative humidity on dried Plantago lanceolata L. leaves during long-term storage: Effects on chemical composition, colour and microbiological quality. *Phytochem Anal.* 2012; 23:88–93.
10. Farkas J. Irradiation as a method for decontaminating food: A review. *Int J Food Microbiol.* 1998; 44:189–204.
11. Beneke CE, Viljoen AM, Hamman JH. Polymeric plant-derived excipients in drug delivery. *Molecules.* 2009; 14:2602–2620.
12. Wilkes, RS. Food compositions incorporating stearidonic acid. *US Pat Appl Publ.* 2009:48.
13. Wolf A, Pouny Y, Marton I, Dgany O, Altman A, Shoseyov O. Use of robust plant chaperonin-like Sp1 proteins to stabilize therapeutic proteins in pharmaceutical use. *PCT Int Appl.* 2007:155.
14. Diederichs J. Plant extract stabilized by polyvinylpyrrolidone and its therapeutic use. *PCT Int Appl.* 2006:13.
15. Miller TJ, Fatnon MJ, Webb SR. Transgenic plants producing soluble, stable immunoprophylactic and therapeutic compositions which released by disrupting plant cell wall. *PCT Int App.* 2004:102.

16. Ludwikowska K, Radecki A, Piekos R. Characteristics of ethoxypolysiloxane oil emulsions. *Akad Med Gdansk Pol Przemysl Chem.* 1969; 48:216–220.
17. World Health Organization (WHO). *Quality control methods for medicinal plant materials.* Geneva, World Health Organization, 1998.
18. Victor CG, Miguel CG, Barbara GS, Martha M. Continuous multi-microencapsulation process for improving the stability and storage life of biologically active ingredients in foods, cosmetics and drugs. *J Microencaps.* 2005; 22:913–919.
19. Stahl CR, Fernandez AO, Woodin FW. Stabilized compositions containing an oxygen-labile active agent. *US Pat Appl Publ.* 2002:6.
20. Iwu M, Okunji C, Tchimene M, Anele N, Chah K, Osonwa U, et al. Stability of cough linctus (Streptol) formulated from named medicinal plant extracts. *Chem Pharm Bull.* 2009; 57:229–232.
21. Miller TJ, Fatnon MJ, Webb SR. Transgenic plants producing soluble, stable immunoprophylactic and therapeutic compositions which released by disrupting plant cell wall. *PCT Int Appl.* 2004:102.
22. Boyong L, Cheng XX. Stable pharmaceutical compositions without a stabilizer. *US Pat App Pub.* 2004:5.
23. Boyong L, Cheng XX. Stable pharmaceutical compositions without a stabilizer. *US Pat Appl Publ.* 2004:5.
24. Guidance for Industry: ICH Q1A (R2) Stability Testing of New Drug Substances and Products. International Conference on Harmonization (ICH) harmonized Tripartite Guidelines, November 2003.
25. Guidance for Industry: ICH Q3A (R) Impurities in New Drug Substances, February 2002.
26. Guidance for Industry: ICH Q6A Specifications: Test Procedures and Acceptance Criteria for New Drug Substances and New Drug Products: Chemical Substances, October 1999.
27. Guidance for Industry: ICH Q1B Photostability Testing of New Drug Substances and Products, November 1996.
28. Guidance for Industry: ICH Q6B Specifications: Test Procedures and Acceptance Criteria for Biotechnological/Biological Products, October 1999.
29. Guidance for Industry: ICH Q1D Bracketing and Matrixing Designs for Stability Testing of New Drug Substances and Products, January 2003.
30. Guidance for Industry: ICH Q1F Stability Data Package for Registration Applications in Climatic Zones III and IV, June 2004.
31. Guidelines for Stability Studies for Human Drugs and Biologics, Center for Drugs and Biologics, Office of Drug Standards, Food and Drug Administration, February 1987.
32. www.ich.org.
33. Nagy KM, Balazs R, Marcisz J, Nagy KW, Tajthy J, Judit E, et al. Plant pigments as light stabilizers for photosensitive drugs. *Ger Offen.* 1992:10.
34. Seca AML, Pinto DCGA. Biological Potential and Medical Use of Secondary Metabolites. *Medicines (Basel).* 2019 Jun 12;6(2):66. doi: 10.3390/medicines6020066. PMID: 31212776; PMCID: PMC6632032.
35. Gamay A. Stabilized vitamin solutions: Use thereof; process for their production; and formulations comprising the same. *US Pat Appl Publ.* 2009:11.
36. Grimm W. Storage conditions for stability testing—long term testing and stress tests. *Drugs Made Ger.* 1985; 28:196–202, (Part I) and 1986; 29:39–47, (Part II).
37. Cho HJ, Won YH, Kim EJ, Choi MH, Jeon IS, Choi WS. Herbal liquid composition capable of effectively preventing precipitation with improved storage stability by comprising β-cyclodextrin, and method for preparing thereof. *Repub Korean Kongkae Taeho Kongbo.* 2007.
38. Shen Z, Zhang L, Yang A. Hair dyes containing Chinese herb impatiens with good safety and stability. *Faming Zhuanli Shenqing Gongkai Shuomingshu.* 2007:17.

16 Regulation of Clinical Trials and the Role of the Institutional Human Ethics Committee

16.1 INTRODUCTION

Ancient civilizations like those of Egypt, Mesopotamia, India, and Greece promote ethics and codes of conduct to be followed by physicians when they are involved in any new findings and research. The ancient history submits the rich ethical inheritance and accountability in patient treatment. Safeguarding the rights of human volunteer participants in any clinical trial, the institutional ethics committee (IEC) or institutional review board (IRB) is formed according to the Declaration of Helsinki and other international guidelines for maintaining the highest integrity in biomedical research. The IEC is not only entrusted with review of proposed research protocols but also is empowered to monitor compliance with all ethical requirements until a study is completed. According to the new modified regulations, the IEC changes composition related to the registration of the clinical trial or function, and cancellation of the registration, suspension of the clinical trial, etc., are well described. In India, the new Drugs and Clinical Trials Rules 2019 (New Rules) entered into force by the government of India on 21 September 2019.

The need for IECs is for safeguarding enforcement of human rights and natural justice in medical and research establishments. The individuals involved in research, such as physicians and research workers, should prioritize ethical standards over personal or corporate interests. This practice ensures that research remains trustworthy, unbiased, and beneficial to society as a whole. Thus, the IEC promotes ethical research and the committee is a combination of multiple constituents and is multi-disciplinary including representation from social work, law, science, academics, healthcare, and others to understand the need for such a study. The IEC represents the patients' and public perspectives and safeguards such that the patients receive maximum benefit.

16.2 EMPOWERMENT OF THE ETHICS COMMITTEE

The Drugs and Cosmetics Rule 122DD empowers the constitution of the IEC with the harmonization of medical, scientific, non-medical, and non-scientific members included for ensuring the rights and safety of human subjects involved in clinical trials. The IEC committee is empowered to review the proposal for research using the

DOI: 10.1201/9781003398431-16

human volunteers to examine how it accounts for investigator suitability, adequacy of methods, facilities, documentation of informed consent, and safeguarding the confidentiality of individuals participating in the study. Also, any adverse drug reaction associated during the trial is to be reported to CDSCO. According to the new drugs and clinical trials rules 2019, the IEC for biomedical and health research is formed under Rule 7 and registered under Rule 8, while the IEC for clinical trials is formed under Rule 16 and registered under Rule 16 of the new clinical trial rules.

Clinical trial rules, 2019 chapter III is directives of the ethical committee as follows:

1. The requirement of the IEC, as per directive of Rule 6.
2. The constitution of IEC for clinical trial as per directive of Rule 7.
3. The registration of IEC for clinical trials, bioavailability (BA), and bioequivalence (BE) studies for clinical trials in accordance with Rule 8.
4. Validity period of registration of IEC for clinical trial, as per the directive of Rule 9.
5. Renewal of registration of IEC for clinical trial, as per the directive of Rule 10.
6. Functions of IEC, as per the directive of Rule 11.
7. Proceedings of IEC for clinical trial, as per the directive of Rule 12.
8. Maintenance of records by IEC for clinical trial, as per the directive of Rule 13.
9. The potential outcomes of a clinical trial's IEC registration include suspension or cancellation done under the directive of Rule 14.

According to the new rules of clinical trial, the IEC has the main two functions. The first function is functioning and approval for bioavailability (BA) and bioequivalence (BE) studies, which described in Chapter III; the second function is IEC responsibilities in biomedical research and health science, which are described in Chapter IV.

16.3 CONSTITUTION OF THE ETHICS COMMITTEE

The IEC is composed of seven members representing the medical, non-medical, scientific, and non-scientific domains: one layperson, one woman, one legal expert, and one independent member from social scientist or representative of non-governmental voluntary agency must be included from social sciences, or representative of a non-governmental voluntary agency must be included. At least 50% of the committee members must be from outside the institute. The chairperson shall be appointed from outside the institute or organization, while the member secretary should be affiliated of the same institute and appointed by a competent authority. The members from the medical sciences should have at least a postgraduate qualification in their specific field of specialization, as well as enough experience and expertise and clarity about their responsibilities and roles in clinical trial.

Every member of the IEC is required to undergo training and development programs, as per CLA directive. The training and development of IEC members are crucial for maintaining the ethical standards and the quality of review processes in research involving human subjects. It helps ensure that IEC members have a strong understanding of their responsibilities and the ability to make informed, ethical, and scientifically sound decisions in their oversight of research studies. Training also contributes to the overall integrity and credibility of the research process. When the trial necessary depends upon the research requirement, the representative from the areas of HIV or genetic disorders, or any specific patient, may be represented in the IEC.

Every IEC member shall declare not to have any conflict of interest, and if any conflict of interest is found, the member may voluntarily withdraw by writing to the IEC chairperson from the IEC review meeting, and such point shall be entered into the minutes in the meeting of the IEC. Also, every IEC member is required to adhere to GCP guidelines for providing rights and safety to the volunteer. Further, the IEC committee may periodically take the action after studying the progress reports of visiting the study sites, including any Serious Adverse Event (SAE) and monitoring internal audit reports and progress reports furnished by the investigator's sponsor. The IEC can reject any proposal as deemed to be unnecessary and reason being given to inform the investigator. Also, the IEC is empowered to discontinue any clinical trial if the study is not clear, if IEC observes any compromise of the rights and safety of volunteers, SAE, etc., and same to be informed by the head of the institution where the clinical trial is being conducted. The IEC is required to study reports of SAE or BA/BE to the CLA for approval—any officer authorized by the CLA to enter the premises where the clinical trial is being conducted is authorized to inspect any records, investigate breaches, or verify the compliance related to the clinical trial. The IEC may constitute more subcommittees for assisting in the functions as deemed necessary by appointing different experts who may not be members. The CLA must be notified within 30 working days of any change in the constitution of the IEC.

16.4 ETHICS COMMITTEE REGISTRATION

Registration of every IEC is required to the Ministry of Health and Family Welfare (MoHFW), Department of Health Research (DHR). The application shall be made under these rules in Form CT-01 to the CLA. Approval of registration if not suspended or revoked, IEC must be provided in Form CT-02 and is valid for five years after the date of issuance. The following content is required to the registration of the IEC as follows.

1. The IEC's composition and constitution:
2. Biodata of all members of the IEC.
3. Standard operating procedures of the IEC.
4. Guidelines followed by IEC as per national or international conventions.
5. All copies related the protocol, case report, investigators' brochures, data collection forms, etc.
6. All correspondence related to application, decision, and follow-up, and to the other members.
7. Minutes and agendas of all IEC meetings with signature of the chairperson.
8. Decisions of every protocol communicated to applicants.

9. Information on any early research termination, including a summary of the causes, must be included.
10. The conclusive report pertaining to the investigation involving digitized microfilms or video recordings.
11. Any recommendation by IEC for compensation of the study.
12. Details of any serious adverse reactions and implemented medical management of trial subjects, and if any compensation was paid.

The CLA registered IEC in Form CT-02 for only BA/BE studies, which is valid for five years from the date of issuance, after carefully reviewing the application by the CLA. If any deficiency is observed, the CLA must return the application within 45 days with the observations in written form, and in such case, an appeal can be made within 60 days.

Application for renewal registration is made by CT-01 with requisite documents between 90 days before registration and the final registration granted by CT-03 for other clinical trial except BA/BE studies. The CLA has power to the suspended or canceled of registration of IEC for biomedical and health research under the rule 18.

16.5 ETHICS COMMITTEE FOR ACADEMIC CLINICAL TRIALS

The Drugs and Clinical Trial Rules, 2019, categorically mention that clinical trials of drugs that have already received approval for a specific use and that are started by researchers, academic institutions, or research centers for new indications, dosages, or dosage forms; the trial's results are only meant to be used for academic or research purposes, not to apply for approval from regulatory bodies the CLA for marketing or commercial purposes.

This type of clinical trial is considered to be an academic clinical trial. The term "academic clinical trial" typically refers to clinical trials conducted primarily by academic or research institutions, such as universities, teaching hospitals, or academic medical centers. While academic clinical trials play a vital role in advancing medical knowledge and patient care, they can indeed have certain limitations compared to trials sponsored by pharmaceutical or industry organizations.

Discussing the roles and responsibilities within the context of academic clinical trials (CT) and biomedical and health research (BHR).

IEC (Independent Ethics Committee): The Independent Ethics Committee, also known as the Institutional Review Board (IRB) in some regions, plays a crucial role in ensuring the protection of the rights, safety, and well-being of human subjects involved in a research study. In the context you've mentioned, the IEC takes responsibility for addressing major adverse events that may occur during the course of the clinical trial.

Compensation for Adverse Events: In clinical trials, adverse events refer to any undesirable experience or side effect that occurs to a participant during the study. Major adverse events are those that are serious and sometimes life-threatening. The IEC is responsible for evaluating the occurrence of such events and determining whether they are related to the study procedures or interventions.

Host Institution's Role: The host institution, where the clinical trial is conducted, is typically responsible for providing compensation or covering the costs associated with compensation. This is often based on the recommendations made by the IEC. Compensation may be provided to participants who experience harm or adverse events as a direct result of their participation in the research.

IEC Recommendations: The IEC reviews and monitors the progress of the clinical trial, assesses the safety of the participants, and makes recommendations to the host institution regarding necessary actions. If compensation is deemed appropriate for participants who experience adverse events, the IEC may recommend the type and amount of compensation to be provided.

Academic clinical studies are considered the investigation is conducted during post-graduation program in the mainstream of the basic research. In the realm of academic clinical trials (CT) and biomedical and health research (BHR), the IEC assumes the responsibility for addressing any instances of major adverse events and the host institution is to provide compensation and/or cover for compensation based on the recommendation of the IEC. ICMR Guidelines are silent about the compensation clause; it is different from the Clinical Trial Rules, 2019.

16.6 TYPES OF ACADEMIC TRIAL

These academic trials play a crucial role in advancing medical knowledge, expanding the evidence base for treatments, and improving patient care. They contribute to the scientific and academic community's understanding of medical interventions and their applications, with an emphasis on non-commercial objectives.

The IEC can refer any cases for clarification to CLA if any doubt overlaps between the academic clinical trials and clinical trials, and within 30 days, the CLA clarify all academic clinical trials to be followed as per the ICMR Guidelines for Biomedical Research on human participants. This process helps ensure that academic clinical trials are conducted in accordance with established ethical and regulatory standards. It allows for guidance and clarification to be provided when there are uncertainties or questions related to the interpretation and application of these standards.

16.7 ETHICS COMMITTEE FOR BIOMEDICAL AND HEALTH RESEARCH

Conducting any biomedical and health research in India, whether by an institute or organization, typically requires approval from an IEC and adherence to the regulations outlined in the National Ethical Guidelines for Biomedical and Health Research involving human participants. These guidelines are a fundamental part of the ethical and regulatory framework for research involving human subjects in India. Such IECs shall be registered using application CT-1 to the Department of Health Research (DHR), MoHFW, government of India (https://naitik.gov.in/DHR/Homepage). After scrutiny of the application, the registration is granted provisionally, which is valid for a period of two years, and final registration is granted via Form CT-03 as earlier mentioned, which is valid for five years until the registration is revoked. Previously, bioavailability and bioequivalence (BA/BE) studies were regulated by the Indian Council of Medical Research (ICMR) guideline called "National Ethical Guideline for Biomedical and Health Research Involving Human Participants" which was released in 2000, amended in 2006 and 2016, and not covered under the Drugs and Cosmetics Rules; the provision of compensation to volunteers is not clearly defined. However, Clinical Trial Rules, 2019 clearly described that research studies encompassing basic, applied, and operational research, as well as clinical research, are primarily conducted to enhance scientific understanding of diseases and conditions, including their detection, causes, and evolving approaches for health promotion, prevention, disease management, and rehabilitation. However, it is important to note that previously these studies do not involve CT (clinical trials) rules. Thus, BA/BE study requires IEC approval, and data of BA/BE study shall be maintained for a period of five years after completion of clinical trial. Rule 16 of the Clinical Trial Rules, 2019 states that the IEC is responsible for compensation and medical management related to any injury or death in a BA/BE study. Other than a BA/BE study, studies such as testing of *in vitro* diagnostics (IVDs), intervention of new surgical instruments, any research on older drugs that is observational and non-interventional, epidemiological health surveys, public health surveys, or user assisted reproductive technology (ART) are concluded in IEC approval.

16.8 LOCAL CLINICAL TRIAL WAIVERS

Any importation of new drugs requires local clinical trial, but under the rule 75 of Clinical Trial Rules, 2019, under specific conditions whereby the local clinical trials are exempted.

Rule 101 of Clinical Trial Rules, 2019, applies if the drug is approved and marketed in certain countries and serious adverse events have been not reported specified by the licensing authorities of those countries. Further, it is deemed and predicted that as it relates to pharmacokinetics (PK) and pharmacodynamics (PD), safety and efficacy will not affect the Indian population and the drugs/molecules are required for emergency treatment of serious diseases, treatment of disease to Indian health scenario, orphan drugs, or the drug is available is not economic or emergency need to India, such drug is permitted directly to trial in a Phase IV study as per design approved by CLA. In such emergency conditions for urgent treatment, and the higher the benefit ratio, the essential studies requirement of other clinical trials like animal toxicology, reproduction, perinatal, teratogenic, mutagenicity, and carcinogenicity are abbreviated, and as per Rule 80 of Clinical Trial Rules, 2019, are allowed directly to Phase IV study if such drugs are marketed in another country for the previous two years. These efforts make new drugs available to India more quickly as compared to global market.

16.9 PERMISSION FOR CLINICAL TRIALS WITH AN UNAPPROVED NEW DRUG

The Drugs and Cosmetic Rules, 1945, specifically Rule 36, provide a regulatory framework in India that governs the granting of licenses for unapproved new drugs intended for individual treatment, especially for life-threatening diseases. The license can be applied for through the Form 12A, along with the prescription of the registered medical practitioner (RMP) from the government hospital, indicating the amount of the drug required. The application can be granted by Form 12B on a priority basis. The Clinical Trial Rules, 2019, clearly indicate that government hospitals are authorized to import or manufacture unapproved drugs in limited quantity for patients suffering from a life-threatening disease or disease may cause permanent disability.

16.10 ROLES OF ETHICS COMMITTEES

The IEC is responsible to protect the dignity, rights, benefits, and well-being of all the human participants in the trial, preserving the principle of justice and ensuring that the benefits of research are fairly distributed among all ethnic groups and all classes in society with the consideration of age, gender, economic status, culture. The IEC should exercise the power to protect and provide safety and protection of rights to all vulnerable subjects in the special populations of hierarchical structure armed forces personnel; medical, nursing, and pharmacy staff and students; prisoners; unemployed or impoverished persons; patients in emergency situations; patients with incurable diseases; ethnic minority groups; homeless persons; refugees; nomads; minors; or others incapable of personally giving consent.

The IEC should complete the review of approved trials of the Institute and ensure that periodic study progress reports should be furnished by the investigators, enforce the investigator brochure when the study protocol and standard operating procedures are mentioned, and ensure that any internal audit shall be submitted by the investigator or sponsor. Any withdrawal of study or stage closure of the research proposal by the IEC and the justification shall be reported to the CLA. The IEC should safeguard the principle of international scientific standards and local community values and customs, be responsive to local healthcare and assist development and education of the research community, and educate the researchers to look after the participants' safety and well-being following each study proposal's scientific validity.

16.11 RECRUITMENT OF RESEARCH PARTICIPANTS

1. A well-defined study design that accounts for the characteristics of the study population, including age, culture, gender, economic status, literacy, and ethnicity, is crucial for conducting research that is scientifically sound and ethically responsible.
2. The characteristics of the population like age, culture, gender, economic status, literacy, and ethnicity are involved in the trial.
3. Predictable risks and inconveniences associated with the anticipated benefits and the risk-to-benefit ratio shall be well explained.
4. The justification of using a control group with correlation of medical ethics shall be explained.
5. Methodology and procedure of logical decision to withdrawing research participants or suspending or terminating the research shall be explained.
6. Provision for the formation of a data safety monitoring committee (DSMC) which can monitor and audit the research, including the participants' safety and health following each study proposal's scientific validity.

16.12 CARE AND PROTECTION OF RESEARCH PARTICIPANTS

Investigator's qualifications and experience related to clinical trial to be considered for the proposed study are as follows.

1. Justification for withdrawal from or withholding of standard therapies for the purpose of the research shall be explained.
2. The provisions of medical care to research participants during and after the course of the research shall be explained.
3. The provisions of medical supervision and psycho-social support to the research participants shall be explained.
4. When any research participants voluntarily withdraw during an experiment, it's important to take specific actions to facilitate their withdrawal and ensure that they have been informed about the study requirements, associated risks, and the consent procedure.
5. Informing the research participants about the requirement of the study, associated risks, and the consent procedure shall be explained.
6. Information to research participants about any financial involvement or cost.
7. Compensations and rewards to the research participants shall be explained.
8. A description of any plans to make the study product available to the research participants following the research shall be explained.
9. Compensation/treatment if associated with any injury disability/death of a research participant shall be explained.
10. The provisions of insurance and indemnity shall be explained.
11. Data confidentiality of the participation shall be explained, including the identity of the patient data, identity, biological samples, and biological samples which shall be maintained.

16.13 THE INFORMED CONSENT PROCESS

Procedure and description of the volunteer recruitment, obtaining informed consent, and the persons involved for obtaining consent shall be explained.

The consent form shall mention the investigation, benefits, and associated risks involved, and the consent form in the direction of adequacy of information, completeness, and understandability is required in the written form with mother tongue or any language, verbal information will be provided to research participants based on their legal representative or guardian, approval if applicable.

If any individual cannot provide consent, the justification shall be provide dand the authorization on behalf of the participations must be recorded in written form with the mention of the relations of such individuals.

Assurances of rights, safety, and well-being of the research participants shall be provided during the entirety of the investigation.

Ensuring the rights, safety, and well-being of research participants throughout the entirety of the investigation is a fundamental ethical principle in research. This includes establishing clear procedures for receiving and addressing queries and complaints from research participants or their representatives.

16.14 THE REVIEW PROCESS

Technical advisory committees (TACs) of the institute are to be informed about the proposal to the IEC before submission.

The investigator must fill up the ethics review application form in simple understandable language with the rational explanation of all points, and submit 12 copies with requisite enclosures to the member secretary of the IEC before a minimum of three weeks before the proposed meeting.

The secretary and member secretary will scrutinize the application, and upon their satisfaction, it will be placed before the meeting.

16.15 RESPONSIBILITIES OF REVIEWERS IN THE IEC INVOLVEMENT REVIEW PROCESS

In the purview of the IEC to review all proposals, including new proposals, revised proposals, amendments if any in the approved proposal, any adverse events reported in the research already going on, and so on. During the meeting, all research proposals are discussed and consultation with the chairperson and member secretary assigned to committee member for rigorous review.

16.16 APPOINTMENT AND RESPONSIBILITIES OF REVIEWERS

The member secretary on behalf of the IEC and on consultation with the chairperson may appoint primary and secondary review. The appointment of primary and secondary reviewers, along with the process of scrutinizing research proposals, addressing changes, re-considerations, and ethical issues during meetings is a standard practice within an IEC to ensure thorough and comprehensive review.

16.17 INVITING RESEARCHERS FOR THE IEC MEETING

The statement highlights an important principle in research ethics—namely, that the ethics review process should be independent and conducted by members of the IEC who have been assigned this responsibility. This independence is crucial to ensure the integrity of the ethical review. The applicant is not required to be present at the meeting because IEC meetings maintain certain confidentiality. Clarification of the submitted protocol may be taken by telephone.

16.18 COMMUNICATING THE DECISION

The member secretary may inform the applicant of the decision. The outcome of the decision, including the suggestion of change of protocol, must be reported and the reason must be provided. Any negative decision shall be supported by clearly defined reasons.

16.18.1 Types of Approval

16.18.1.1 Straight Approval or Approval with Comment

The proposal is granted when the IEC has no questions. Sometimes, IEC members put comments on the approved proposal or recommendations for the future submission of the research proposal, and such comments are included in the approval letter.

16.18.1.2 Conditional Approval

Conditional approval is a mechanism to help researchers meet ethical and legal requirements, ensuring the safety and well-being of research participants. It also demonstrates the commitment of the IEC to work with investigators to facilitate responsible research practices. The investigator addresses the raised comments and compiles all the comments and submits them to the member secretary. The compiled answers shall be vetted by all IEC members and the concurrence of the chairperson the approval shall be granted.

16.18.1.3 Contingent Approval

This type of approval study could not be initiated until being cleared or approved by IEC. In principle, the IEC members are agreed on the protocol but some concern arises about the study protocol example, the protocol, the recruitment, and/or the consent form. In such cases, the investigator replies to the chairperson, whose review may or may not conclude with a decision and further referral to the IEC. In such conditions, no additional issues or deficiency is pointed out unless the government/legal requirements are ignored during the meeting, which may be serious, or in the opinion of the chairperson, any serious point was inadvertently failed to be noticed during the review. The approval or approval number of endorsed until the investigator satisfactorily acknowledged the raised issues.

16.18.1.4 Returned for Additional Information

The IEC is not permitted to follow for further consideration of the proposal until submission of required additional information and further review. In the case of deficiency like risk-to-benefit ratio of the trial, participant protection, ethical issues, or a non-convincing application, in such cases, the clarification or additional information is required for approval of the proposal. The investigator replies in writing in a timely manner all the issues raised by the IEC or subcommittee and the right of decision is reserved to the members or the chair. If the IEC is content with the raised question addressed in the revised proposal, when the matter comes up for a second review, it could be approved unconditionally, conditionally, or differently. However, the IEC is authorized to return back again to the investigator if the proposal fails to address particular issues.

16.18.1.5 Denial of Approval

The denial of a proposal is framed on the basis of their existing several irregularities; the risk is higher compared to potential benefit, inadequate pre-data validation, violation of the ethical norms of the participations, safety standard of the participations, etc. It could be that the IEC requested a reply from the investigator and even after the reply, questions were raised which mean that the query was unsatisfactory addressed, and then the IEC declined to approve the proposal. The IEC must invite the investigator and explain the reason of denial of the project then provide opportunity for the investigator to present their views and an explanation. The denial letter should contain the reason for the denial of the approval of the study proposal.

16.18.1.6 Reversing Positive Decision

The IEC is authorized to reverse a positive decision if any study is found to be adversely affected during the clinical trial and the benefit-to-risk ratio changes sufficiently.

16.18.1.7 Withdrawing a Proposal/Application from Review Process

The investigator is authorized to withdraw the proposal if deemed necessary and the application shall be submitted to the member secretary and clearly state the withdrawal of the the application prior to the IEC meeting.

16.18.1.8 Discontinuation of Trial

The application from all the research proposal irrespectively mentions about the discontinuation of the proposal of the study and accordingly the IEC approved the study. Any research proposal may be withdraw in midway when the such overlaid conditions of withdraw is violated and such cases the notify to investigator to stop the trial another ground of the discontinuing of trial is that achievement of the goals is not reflected to the trial or the result conclusion is not favor the hypothesis or unambiguously or disproving hypothesis. The IEC shall be notified of any prematurely terminated study, along with reasons for termination and a summary of the results of research conducted.

16.19 EXCLUSION OF ETHICS REVIEW

Certain studies are exempted from the IEC even though they involve human subjects, as follow.

1. Research involved with certain established normal educational settings and special educational instruction strategies, or the effectiveness and comparison of the different instruction techniques, classroom management or curricula, or methods, are exempt.
2. Research involving any educational tests or public behavior, such as cognitive, aptitude, diagnostic, achievement, or survey or interview procedures, are exempt.
3. The IEC plays a critical role in ensuring the protection of research participants and the ethical conduct of research. When considering cases where information or interviews are recorded and later may be identified, the IEC should take into account various factors related to confidentiality, privacy, and potential harm to research subjects. Further, any disclosure of the human subjects may breach confidentiality or attract any civil or criminal liabilities or damaging individual reputation, employability and financial standing.

4. Any existing data, records, documents, and pathological specimens or diagnostic specimens whereby no approval of IEC is required but the IEC's clearance is either directly or indirectly necessary if the investigator records the data in a way that makes subjects identifiable.
5. There is no exemption if it involves children and other vulnerable groups as participants.

16.20 PROCEDURE FOR OBTAINING EXCLUSIVITY OF ETHICS REVIEW

The investigator is required to submit a request application for the exemption of review to the technical advisory committee with the abstract of the proposed research proposal, and the same application is forwarded to the chairperson of the IEC. The chairperson expedites the review process as to whether the application has merit for consideration for exemption. If the research proposal establishes that IEC approval is not required, this determination is to be provided to the investigator by the member secretary at the next convened meeting of the IEC. Otherwise, the investigator is informed accordingly and submits the research proposal to the IEC meeting.

16.21 EMERGENCY OR INTERIM MEETINGS

Emergency or interim IEC meetings permit only the revised decision in certain serious adverse cases reported or if any adverse events are reported during the clinical trial. The chairperson is authorized to constitute any emergency or interim IEC meeting, and any decision at an interim meeting is endorsed in the next meeting. Emergency or interim IEC meetings are a critical response to the urgent research needs that arise during pandemic situations. These meetings facilitate the rapid review of research proposals while upholding ethical standards and ensuring the protection of research participants. Such flexibility in the IEC review process is essential to address public health crises effectively.

16.22 EXPEDITED REVIEW

A research proposal can be considered for expedited review if it has minimal risk or no adverse reaction is expected and such study is based on daily life whereby certain interventions not more than psychological or physical examination help to access or improve the quality of life. In the expedited review process, the full IEC is not required and the chairperson is authorized to conduct the meeting by involving the IEC members or a subcommittee. The main limitation of the expedited review process is that no vulnerable or special category population or any genetic study or phylogenetic study are to be included; in such a case, the proposal is to be considered in a normal meeting. If the investigator has the opinion that the proposal may be taken in expedited review, they can apply to the technical advisory committee or competent authority and also forward to the chairperson of the IEC. The chairperson evaluates the merits of proposal and determines the expedited review quality, and accordingly takes the necessary actions. The committee member is authorized to modify/revise the submitted proposal, except its rejection, and inform the investigator of the same. If the proposal is not considered for approval, then this proposal is placed to the next IEC meeting. The approved expedited studies are conveyed to the next IEC meeting and enlisted as the minimal risk group of study.

16.23 POST-AMENDMENT OF APPROVED STUDY

The approved study is always monitored by IEC and a minimum of one review each year is required for continuing; this provision is applicable to all the studies which are not closed and the studies in which data has been collected or those which involve data analysis. There is a provision that any changes required are deemed by the investigator and a request to submit such changes is to be made to the technical advisory committee or competent authority, which may forward to IEC with recommendation to IEC for further consideration. After the evaluation of such changes, the member secretary may issue an amendment letter.

If changes the recruitment process, study design, volunteer criteria, change of place of study, consent process, inclusion of new methods, or any instruments or any new parameters, happen during a trial such changes are considered major and these are mandatory to be submitted the IEC. After evaluation and adhering to strict compliance of regulations, the IEC may approve and issue an amendment or post-approval letter. However, any other changes in the post-approval study may be considered minor changes or no changes in the approved appraisal, which can be considered for the expedited review. Any adverse reaction or immediate hazards found to the research subjects may change the protocol without IEC approval and must be reported to the IEC and submitted with all supporting documentation for the next convened meeting. When an investigator identifies an issue with the risk/benefit assessment in an approved research protocol, they have an ethical obligation to inform the IEC and request an amendment to rectify the inaccuracies. The collaboration between the investigator and the IEC is essential to ensure the research is conducted ethically and in compliance with regulatory standards. The principal investigator is responsible for identifying and requesting amendments to an approved research protocol and for conducting an initial evaluation of the proposal. Collaboration with the IEC is crucial to achieve these goals. Additionally, adhering to specified time frames is important to ensure timely research conduct and maintain ethical and regulatory compliance. Delays in the evaluation process may lead to actions such as the termination of the research proposal.

16.24 RECORD RESTORATION

As per the ICMR directives, the institute casts the following information to the public domain.

1. Constitution and composition of the IEC members and their curriculum vitae.
2. SOPs and quality policy.

3. International, national and local guidelines adopted by IEC for review.
4. All forms prescribed by the IEC and its annual calendar of meetings.
5. Training/education on research bioethics shall be conducted by the IEC.
6. Annual or biannual reports and audit reports of IEC.

16.25 RESTORATION BY THE INSTITUTE

The member secretary is the custodian of all IEC -related documents, including the applications, agendas, records, notifications, communications to the IEC members, minutes, and all protocols, etc. The relevant records of the IEC shall be maintained for a minimum 10 years, except any litigation, in which case, such documents are stored until the litigation is resolved. Protocols are maintained according to the field and approved proposals, along with the all written documents. Tracking and confidentiality are to be adhered to during any access of data and retrieval procedures, and such data shall be systematically stored in electronic form.

16.26 MINUTE PREPARATION, APPROVAL OF IEC, AND MONITORING OF THE RESEARCH AND ITS ETHICAL CONDUCT

The meetings including the approval of any research proposal are included in the minutes and the member secretary prepares the minutes. The minutes of meeting are circulated to all IEC members and accorded approval by the chairperson. Usually, the minutes of the meeting are read out in the next meeting of the IEC, discussed, and conclusive with or without amendments. The certificate is issued to the investigator after granting of approval for the proposal, and the member secretary ensures that the approved protocol is indexed for the archives. The process involves reviewing the minutes, discussing any necessary changes, and then reaching a conclusion on whether to accept the minutes as they are or with the suggested amendments. Once the minutes are deemed conclusive, they are considered an accurate record of the proceedings. This process ensures transparency and accuracy in documenting the decisions and discussions of the Institutional Ethics Committee.

The approved protocol provides a legal right to the biomedical research involving human subjects, while perusing the approval procedure thoroughly, the research proposal must be investigated to ensure legal compliance as it raises future questions about the safety and involvement of certain venerable populations like pregnant and lactating women, children, vulnerable subjects, and those with diminished autonomy, research population groups like defense personnel, ethnic populations, and the legality of any research issues pertaining to commercialization and international collaboration.

Further, the IEC must take precautions before any approval of the study and must scrutinize any following issues.

16.27 CONSIDERATION OF INSTITUTIONAL ETHICAL PERMISSION OF DRUGS

Most drugs or pharmaceutical formulations, new combinations, different claims of the existing molecules, or new molecules are to be evaluated according to the new

Clinical Trial Rules, 2019. The recruited volunteers, prerequisite requirement data, etc., are to be confirmed before approval.

16.28 CONSIDERATION OF INSTITUTIONAL ETHICAL PERMISSION FOR BIOMEDICAL DEVICES

In clinical trials involving human volunteers, especially those that evaluate biomedical devices, it's crucial to understand and consider the classification of the device according to relevant regulations (WHO, USFDA, or Medical Device Rules 2017), as well as assess associated risks. Additionally, the legal status of the biomedical device and the availability of standard devices are significant factors. Understanding the classification of biomedical devices, associated risks, legal status, and the availability of standard devices is integral to the ethical and safe conduct of clinical trials involving human volunteers. Regulatory compliance, ethical review, and transparent communication with participants are key elements in the process. It must be addressed clearly whether the proposed trial includes invasive or non-invasive processes. The IEC must exercise prudent judgement when conducting any study involving radiological investigations, such as magnetic resonance imaging, thermography, electrocardiography, electroencephalography, electroretinography, detection of naturally occurring radioactivity, diagnostic infrared imaging, ultrasound, doppler blood flow, or echocardiography—require understanding of the purposes of the study whether they are solely for non-research or research purposes such as medical treatment or diagnosis. Unnecessary exposure to radiological examination cannot be endorsed by the IEC, and the biomedical and clinical safety shall be well established before exposure by any new machine. Routine purposes used for any radiological examination require regulatory clearances and instruments must be approved in the market. Expedited approval of any radiological examination IEC is not possible.

16.29 CONSIDERATION OF INSTITUTIONAL ETHICAL PERMISSION FOR NON-INVASIVE WITHDRAWAL OF BIOLOGICAL SAMPLES

The technical competence of the IEC is vital for evaluating the safety and ethics of research procedures, such as blood withdrawal from volunteers. The specific guidelines provided in the statement are important for ensuring participant safety and ethical compliance. IEC members should be well-equipped to make informed decisions that prioritize the welfare of research participants. The proposal must mention the body weight, age, and volume of blood withdrawn in the time schedules. Ideally more than 50 kg adult and blood withdraw volumes 500 ml in 8–10 weeks with maximum two times in the week is permitted.

Samples collected from volunteers by non-invasive processes and carefully examined include amniotic fluid only obtained at the labor or the delivery or deciduous teeth or permanent teeth during the process of shedding or when extraction is necessary for medical purposes. Numerous other traditional instances can be made simpler, such as the requirement that saliva be collected solely using an accumulated technique either or by simulation by chewing wax or gum base or by using the diluting effect of citric acid within tongue. The collection of hair and nail samples is permitted by the clipping method, and likewise, skin or mucosal cells shall be collected by buccal scraping or swab, skin swab, or mouth washings.

Research perception, identity, motivation, cognitive processes, language, communication, cultural practices or beliefs, and social behavior or research using survey, interview, or oral history methods of the individual or group study program evaluation, human factors evaluation, or quality assurance methodologies require clearance from the IEC and approval shall be judiciously given without violating natural justice or certain democratic privileges granted by the constitution or laws.

16.30 VOLUNTEER CONFIDENTIALITY

The safety and privacy of volunteers shall be protected by the IEC at the time of collection of data from voice, digital, video, or image recordings made for research purposes. The IEC also ensures in protocols that data documents, records, or specimens which have been collected from volunteers cannot reflect their identity or social lives.

16.31 MISCELLANEOUS

At the time of approval of any study consider the purposes of the study and whether the study is associated with academic or commercial application. Further, also require understanding whether the study is previously approved or the project is permanently closed; whether any adverse effects is reported related to the project or no volunteer are recruited and are permanently ceased or the long time follow up require to complete the research work. The research proposal is also understood if any additional risks are involved or the research activities are only limited to the data integration and such matter IEC shall be acknowledged and resolved in the next convened meeting.

The IEC's role is to comprehensively evaluate research proposals, taking into account the study's purposes, history, recruitment status, and potential long-term follow-up needs. This evaluation ensures that the research is conducted ethically, with a focus on participant safety and welfare.

16.32 ADVERSE REACTION REPORTING AND MANAGEMENT

The proactive identification and reporting of anticipated adverse reactions during a clinical trial are critical for ensuring participant safety and the overall integrity of the research. The investigator shall report within 24 hours to the IEC members or the Data and Safety Monitoring Board (DSMB), and any delay in this reporting must be explained. The multicentric trial DSMB will immediately stop the trial and the investigator is asked about the reason for the adverse reaction and whether the adverse reaction is related to the trial. The adverse reactions are two types which are serious adverse and second are the serious adverse reports that lead to death.

16.32.1 Serious Adverse Reactions Other than Death

This type of adverse reaction may cause permanent disabilities. The investigator may change the protocol of the investigation, informed consent, or management of the research proposal. The IEC is empowered to visit the trial site to investigate actual conduct and/or by monitoring DSMB as a necessary step for ensuring that research proposal is followed according to the approved study by IEC. Any serious adverse effects reported shall be analyzed by the investigator and the sponsor, followed by forwarding to the CLA, the chairperson of the IEC, and the head of the Institute within 14 days. The opinion is persevered about financial compensation and shall be provided within 30 days by the sponsor as per the Seventh Schedule of the Clinical Trial Rules, 2019.

CLA is fully authorized to take the opinion by appointing an expert committee to determine the cause of SAE and judiciously decide on compensation within 60 days of receiving a report of SAE. Further, the IEC must take the extra attention to the BA/BE study, whereby also the compensation is provisioned any injury or disabilities and shall be paid according the Seventh Schedule of the Clinical Trial Rules, 2019 within 90 days of the report of SAE.

16.32.2 Serious Adverse Reactions and Death

The prompt reporting of Serious Adverse Events (SAEs) in clinical trials is indeed a critical and essential aspect of ensuring participant safety and maintaining the integrity of the research process. This critical event of any trial and such serious adverse effects (SAE) must be reported by the sponsor or investigator within 14 days to CLA and the chairperson IEC and head of the Institute. The analysis on SAEs and the determination of compensation, especially in cases of death, should follow a specific formula as specified in the Seventh Schedule, Clinical Trial Rules, 2019 to be sent to the CLA and required to pay compensation by the sponsor or a designated representative within 30 days of the investigator's report on the SAE of death. The IEC member has the responsibility to collect the relevant documents from the investigator or sponsor and wisely forward all documents vetted by the chairperson to arrive at the conclusion about the CT-related death. The appointed expert committee forwards the reports to the CLA within 60 days from the first receipt of the report of the SAE and validates the reports of the IEC, investigator, and sponsor.

Adhering to these rules and regulations is critical for ensuring the ethical conduct of clinical trials and the protection of the rights and well-being of research participants. Timely and appropriate compensation for SAEs is a fundamental component of participant safety and welfare in clinical research. Similarly, the CT or the BA/BE study is to be followed and the quantum of the compensation shall be paid as per the Seventh Schedule, Clinical Trial Rules, 2019 within 90 days of SAE. The investigator or sponsor shall pay all reimbursement within 30 days after getting the order from the CLA.

The term "quantum of compensation" refers to the amount or extent of compensation that will be provided in the context of a clinical trial or a bioavailability/bioequivalence (BA/BE) study. In other words, it specifies the measure or degree of financial or other forms of compensation that will be awarded to participants who experience a Serious Adverse Event (SAE) during the study.

The "Seventh Schedule, Clinical Trial Rules, 2019" likely contains a structured framework or guidelines that outline how compensation should be calculated and awarded based on the nature and severity of the adverse event. This schedule may take into account various factors such as the type and extent of harm caused, medical expenses incurred, and the impact of the adverse event on the participant's well-being.

The inclusion of a specific timeframe, "within 90 days of SAE," indicates the urgency and importance of providing compensation promptly to participants who have experienced serious adverse events during the course of the clinical trial or BA/BE study.

It's essential for clinical trials to have clear and ethical guidelines regarding compensation to ensure the well-being and rights of participants. These guidelines are typically established by regulatory authorities to maintain ethical standards and promote the safety of individuals involved in research studies.

16.33 COMMENSURATION RULES

Clinical Trial Rules, 2019 are enforced to ensure that all compensation will be provided to the trial subjects as Rule 42, whereby it is categorically directed that before the trial investigator and sponsor shall declare that all compensation, including the death or permanent disability or any injury, is deemed as serious adverse reaction. The financial compensation for death or considering the major injury or permanent disability is calculated excluding the expenses of medical management. The expenses of medical management shall be paid to the trial subject additionally, along with compensation.

The injury is not treated as serious during the trial and in such cases the investigator or sponsor shall compensate for loss of wages. In the BA/BE study, sponsor will incur all the expenses in medical management as long as required for any injury related to CT. For any injury or death or permanent disabilities associated with BA/BE study, the compensation rule 42 of Clinical Trial Rules, 2019, shall be applicable.

Deviations from the agreed-upon compensation terms in a Bioavailability (BA) or Bioequivalence (BE) study are indeed considered serious offenses and can lead to various consequences, including the termination of the study and restrictions on future involvement in research. These measures are in place to uphold the highest ethical and scientific standards in clinical research, especially in studies involving investigational drugs or medical products. Ensuring that compensation agreements are honored and that investigators and sponsors comply with all relevant regulations and ethical guidelines is essential for maintaining the credibility and integrity of the research process.

The genesis of adverse effects is carefully scrutinized in clinical trials, and any violations of protocol, bioethics, or scientific misconduct can cast doubt on the reliability of the data and the safety of the investigational product. It is the responsibility of research oversight bodies, regulatory agencies, and ethics committees to thoroughly investigate and address these issues to ensure the protection of human subjects and the integrity of the research process. Further, any concomitant medication which was not considered in approved protocol or deviation of any standard care or rescue medication are also considered the cause of SAE. The improper blank study without standard care, child involvement in any trial without parent, or any misleading CT procedure also attracts the SAE.

16.34 CLAUSE OF CONFLICT OF INTEREST AND BREACH OF TRUST OF IEC

It is understood that all IEC members maintain confidentiality about the study, results, and participation with highest level. All IEC members sign an agreement which addresses this concern. Any issues arising directly or indirectly which attract conflict of interest the IEC member must be reported immediately. Any proposal submitted by any IEC member, or one in which any IEC member is directly or indirectly involved, is considered a conflict of interest and that IEC member is not to be allowed to participate in the meeting when the proposal is discussed or any events when the proposal is decided.

16.35 ACCESS TO DRUGS BY THE PARTICIPANT AFTER THE TRIAL

After a successful clinical trial, the investigational drugs or new drugs may be provided to a trial subject if no alternative to the treatment is available and such trial drugs were proven to possess beneficial effects to the trial subject; in such cases, even after scheduled trial, the drug may be provided to the trial subject. In such cases, the trial subject or legal heir has to consent to use of the drug without any legal liability of the sponsor or investigator if any SAE are observed.

16.36 REVIEW AND AUDIT OF THE WORK OF IEC

It is one of the important assignments to the member secretary to ensure GCP by reviewing the past proposals at regular intervals, and important findings of any study may be published to the institutional website. Also, it is important to publish the annual report of the clinical trial conducted, along with an abstract of the results. The member secretary shall audit all the approved clinical trials annually and the reports shall be maintained for future reference.

BIBLIOGRAPHY

1. G.S.R.227 (E): New Drugs and Clinical Trials Rules. 2019. Ministry of Health and Family Welfare, India. March. Available from: https://cdsco.gov.in/opencms/opencms/en/Notifications/Gazette-Notifications/. Accessed 5 August, 2020.
2. ICMR. 2016. National Ethical Guidelines for Biomedical and Health Research Involving Human Participants. 12 October. Available from: www.icmr.nic.in/guidelines/ICMR_Ethical_Guidelines_2016.pdf. Accessed 5 August, 2020.
3. Jain P., Chauhan R. 2019. India's new drugs and clinical trials rules: An industry perspective. *Regulatory Focus*. Available from: www.raps.org/news-and-articles/news-articles/2019/7/indias-new-drugs-and-clinical-trials-rules-an-in. Accessed 1 August, 2020.
4. Singh N., Nivedita JM., Gokhale PM., Parmar DV. 2020. New drugs and clinical trials rules 2019: Changes in responsibilities of the ethics committee. *Perspect Clin Res*. 11: 37–43.

Index

For Product Safety Concerns and Information please contact our EU representative GPSR@taylorandfrancis.com
Taylor & Francis Verlag GmbH, Kaufingerstraße 24, 80331 München, Germany

www.ingramcontent.com/pod-product-compliance
Lightning Source LLC
LaVergne TN
LVHW010551110826
845149LV00003B/621

* 9 7 8 1 0 3 2 5 0 4 3 0 8 *